High-Energy Batteries

High-Energy Batteries

Raymond Jasinski
Tyco Laboratories, Inc.
Bear Hill
Waltham, Massachusetts

℔ PLENUM PRESS · NEW YORK · 1967

Library of Congress Catalog Card No. 66-22124

© 1967 Plenum Press
A Division of Plenum Publishing Corporation
227 West 17 Street, New York, N. Y. 10011
All rights reserved

Preface

Many of the present military, space, and commercial requirements for packaged electrical power are exceeding the capabilities of traditional battery systems, designs, and configurations. As a result, there has been renewed interest in the subject of electrochemical energy storage and energy conversion, in particular high-energy batteries and fuel cells. This interest is reflected by the increase in the number of publications on batteries and related electrochemistry in recent years, as the data listed below indicate.

	Number of papers (open literature)	Number of papers (Russian literature)
1945–1949	133	38
1950–1954	208	48
1955–1959	494	212
1960–1964	684	282

This volume is intended to be a summary of the basic principles and technology of battery performance, with the emphasis placed on high-energy batteries. It is not intended to be a compilation of individual recipes employed by manufacturers in constructing working devices. Recently, a number of books have appeared on the subject of fuel cells, so that this topic need not be considered here.

The basic sources of information considered were as follows: (1) the open literature; (2) government reports, principally those of the Department of Defense and NASA; and (3) the patent literature. For convenience, references to the latter source are listed separately; in the text, these references are designated by the letter "P."

The information surveyed covers the time period from 1945 to the present. Two excellent books by Dr. G. Vinal, then of the National Bureau of Standards, describe the pertinent developments made prior to 1945.

v

TABLE I

Factors Controlling Battery Performance M_w

Figure-of-merit term	Positive plate	Interface	Electrolyte	Interface	Negative plate	Accessories
Q_0	1. Plate loading	1. Oxidation potential 2. Rate of electron transfer 3. Number of electrons transferred per molecule	1. Decomposition potential 2. Electrolyte stoichiometry	1. Reduction potential 2. Rate of electron transfer 3. Number of electrons transferred per molecule	1. Plate loading	
K_1	2. Availability of active material for discharge: pore structure, surface area, thickness of plate, solubility, forming procedures 3. Conductivity of plate, depolarizer, binder; variations in conductivity with dimensions of plate and progress of discharge	4. Mass transport of reactants into and out of the interface, through a film of product	3. Mass transport across electrolyte; concentration gradients 4. Ionic conductivity and variation with the progress of discharge	4. Mass transport of reactants into and out of bulk electrolyte, through a film of product, soluble or insoluble	2. Availability of active material for discharge (see item 2, positives) 3. Conductivity of the plate and the variations with plate dimensions and progress of discharge	

(negative electrode)		(separator / electrolyte)	(positive electrode)		(battery system)
4. Changes in physical properties on wet stand, e.g., loss of surface area from recrystallization 5. Spontaneous decomposition on dry stand; loss of adherence to grid 6. Thermal stability 7. Effect of temperature on reaction kinetics 8. Adsorption of impurities onto surface from electrolyte 9. Surface morphology	5. Corrosion and solubility 6. Oxidation of separator 7. Effect of temperature on reaction kinetics 8. Adsorption of impurities onto surface from electrolyte 9. Surface morphology	5. Chemical stability of the separator 6. Degradation on cycling 7. Thermal stability of solvent, solute, and separator 8. Effect of temperature 9. Ability of separator systems to prevent diffusion of reactants and products out of electrode cavities 10. Weight and volume of electrolyte	5. Corrosion and solubility 6. Penetration of separator by dendrites 7. Effect of temperature on reaction kinetics 8. Adsorption of impurities onto surface from electrolyte 9. Surface morphology	4. Changes in plate on wet stand and operation, e.g., dendrite growth and recrystallization 5. Oxidation on dry stand; loss of adherence to grid 6. Thermal stability of active material	1. Weight of container, leads 2. Activating mechanisms for reserve batteries 3. Gas recombination or venting 4. Sealing mechanisms 5. Recharging equipment 6. Heat sinks, radiators, heaters
K_2 7. Weight and extent of grid and support structure; quantity of conductive binder to provide mechanical stability and minimal ohmic loss				7. Weight and extent of grid or support structure for mechanical stability and minimal ohmic loss	

Within the context of this volume, the term high-energy batteries applies to a minimum energy density of at least 100 W-hr/lb for primary batteries and 50 W-hr/lb for secondary batteries. The term primary battery is used in its most general sense, including reserve and nonreserve devices. It is realized that one single parameter will not be adequate for comparing and evaluating all battery systems under all duty regimes. However, energy–weight and energy–volume densities are of dominant importance for the majority of commercial and military applications which have arisen in recent years.

As shown in Table I, the energy–weight density of a battery system is dependent upon the combination of a large number of physical and chemical factors. It was therefore necessary to form a precise outline by which the large amount of available information could be organized and displayed. It is proposed that the energy density of a battery be represented as the product of three terms as follows:

$$M_w = Q_0 K_1 K_2$$

where M_w is the observed energy density, in units of watt-hours per pound of total battery; Q_0 is the calculated thermodynamic energy density of the redox couples in a given battery; K_1 is a dimensionless quantity representing the inefficiency of the battery in converting all the active material to product at the theoretical potential; and the "weight efficiency" K_2 is defined as $w_0/\sum W$, where w_0 is the weight of the redox couples and $\sum W$ is the weight of the entire system. By analogy with other fields of energy conversion, M_w is called the figure of merit. In this general form, the figure-of-merit expression serves as the organizational outline for correlating and displaying the large body of experimental and descriptive information available on batteries.

The material in this book is, therefore, arranged as follows. The first chapter is a discussion of the basic electrochemistry of the battery discharge. The general polarization effects, parasitic effects, and weight factors which determine battery performance (M_w) are indicated in Chapter 1. The capabilities (Q_0), advantages, and limitations of specific plate materials, solutes, and solvents are treated in Chapters 2, 3, and 4. Chapter 5 is concerned primarily with the specific parasitic processes which inhibit the complete consumption of active material (K_1). In Chapter 6, the structural features of batteries are discussed (K_2). For convenience, battery charging is discussed separately (Chapter 7). The final chapter is a brief summary of the state of the art of selected high-energy systems.

This book is an outgrowth of a study of high-energy batteries carried out for the U. S. Navy, Bureau of Weapons, under Contract NOw-64-0653f. The contract was initiated and supervised by Mr. Milton Knight of the Research Division, Bureau of Weapons, whose suggestions and advice have been of great value. I also wish to acknowledge the helpful comments of Mr. S. R. Marcus and Mr. B. E. Drimmer of the Bureau of Weapons; of Dr. Robert Hamilton of the Institute for Defense Analysis; and of Dr. A. G. Hellfritzsch of the U. S. Naval Ordnance Laboratory.

I am indebted to my colleagues at Tyco Laboratories, Inc., for their advice and assistance in the project. I especially want to acknowledge the assistance of Mrs. Sharon Cooper in assembling the input literature and of Dr. A. C. Makrides, Dr. J. N. Butler, Dr. S. B. Brummer, and Mr. M. Ruben for their helpful criticism on the content and organization of the material.

Raymond J. Jasinski

October 25, 1966
Waltham, Massachusetts

Contents

Chapter 3

Electrochemically Active Materials: Nonaqueous, Inorganic Electrolyte Systems

Chapter 8
State of the Art—Performance

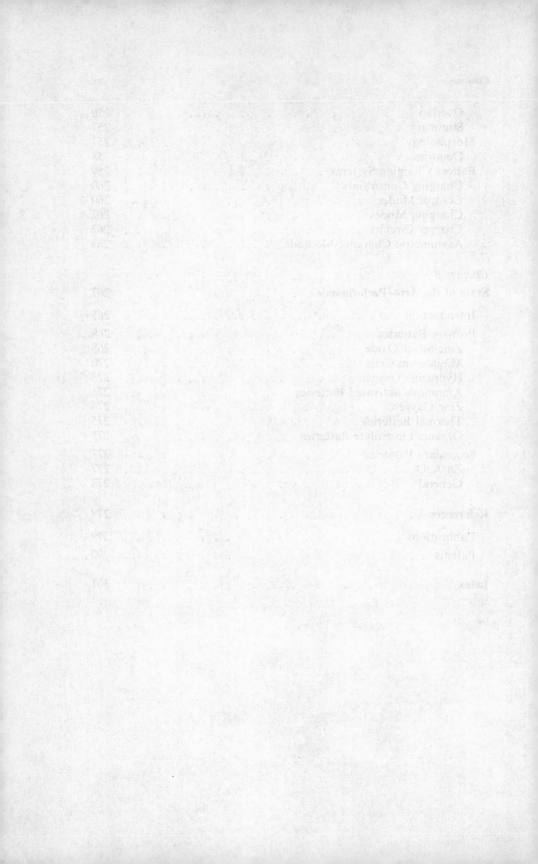

Chapter 1

Electrochemistry of Battery Discharge

INTRODUCTION

This chapter is concerned with the state of the art of the theory describing the general electrochemical processes in batteries. The processes considered are those phenomena which take place at the electrode–electrolyte interface and those within the bulk electrolyte. The equations for simple electrode processes are developed first; the complications introduced by the use of a porous structure are then indicated. Since much of today's research and development is concerned with solvents other than water, a brief treatment of the specific role of the electrolyte in the plate discharge and in ion transport is given.

The discharge of active material is often accompanied by parasitic processes, such as corrosion and dissolution. These are discussed in general; specific examples are given in Chapter 5.

ELECTROCHEMICAL EFFICIENCY

Basic Equations

Consider a first-order charge transfer reaction of the following type occurring on an electrode of constant surface area:

$$M^{+n} + ne^- \underset{a}{\overset{c}{\rightleftharpoons}} M$$

It has been shown [A 297,298] that the removal of bare, unsolvated cations from their energy wells in a metal lattice requires several electron-volts. The free energy change accompanying the solvation of ions is also equivalent to several electron-volts, so that the overall

1

energy change can be relatively small. It is possible to visualize an
energy well for a cation in solution separated by an activation-energy
barrier from the energy well in the metal lattice. This is illustrated
schematically in Fig. 1-1. In terms of transition-state theory, the net
rate of reaction i is given by the rate of reaction from solution to metal
minus the rate of reaction from metal to solution

$$i = i_c - i_a \qquad (1\text{-}1)$$

where i_c is the cathodic current and i_a is the anodic current.

In the absence of electrical potential effects [Fig. 1-1(a)], this net
rate is given by an equation of the form

$$i = X_1 \exp\left(\frac{-\Delta G_c{}^*}{RT}\right) - X_2 \exp\left(\frac{-\Delta G_a{}^*}{RT}\right) \qquad (1\text{-}2)$$

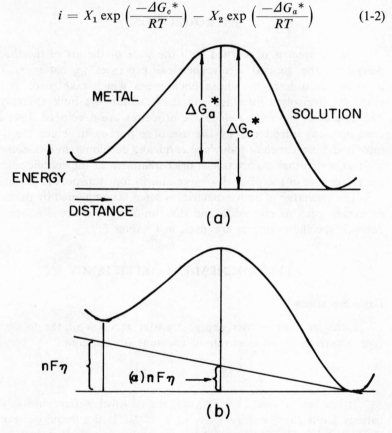

Fig. 1-1. Energy–distance relationships at the metal–solution interface [A 297].
(a) Equilibrium.   (b) Anodic process.

where X_1 and X_2 are parameters involving the concentrations of the reactants.

The effect of electrical potential is illustrated in Fig. 1-1b, where a fraction of the potential aids the reaction from left to right and the remainder aids the reverse reaction. In terms of the anodic and cathodic currents, this potential dependence is described as follows:

$$i_c = nFk_1C_{M^+} \exp\left(\frac{-\Delta G_c^*}{RT}\right) \exp\left(\frac{-\alpha nF\Delta\varphi}{RT}\right) \qquad (1\text{-}3a)$$

and

$$i_a = nFk_2C_M \exp\left(\frac{-\Delta G_a^*}{RT}\right) \exp\left[(1-\alpha)nF\frac{\Delta\varphi}{RT}\right] \qquad (1\text{-}3b)$$

where C_M is the concentration of metal atoms in the metal interphase (metal surface); C_{M^+} is the concentration of metal ions in the solution interphase; α is the charge transfer coefficient aiding the cathodic reaction $(0 < \alpha < 1)$; k_1 and k_2 are rate constants; and F is the Faraday constant. The sign of $\Delta\varphi$ is established by use of the international (Stockholm) convention. The term $\Delta\varphi$ is the potential difference across the Helmholtz double layer, and it is this potential which influences the rate of electron transfer. In concentrated electrolyte, the diffuse double layer is effectively collapsed, and $\Delta\varphi$ is the potential drop from the electrode surface into the bulk of solution. It must be emphasized that this simplification is not valid for dilute electrolytes, and experiments on such systems must consider this double-layer effect. If the discharge step corresponds to part of a sequence of consecutive reactions in an electrochemical process and must occur v times in order to produce the appropriate number of gram-atoms or radicals for formation of one mole of product, then it is appropriate to include the stoichiometric number v in equations (1–3a) and (1–3b). The term v will appear in place of n as N/v if N is the total number of electrons involved in one act of the overall reaction; n is the number of electrons involved in the discharge step [A 377].

For convenience, we will make the following definitions:

$$\Phi = RT/F$$

$$k^+ = k_1 \exp\left(\frac{-\Delta G_c^*}{RT}\right)$$

$$k^- = k_2 \exp\left(\frac{-\Delta G_a^*}{RT}\right)$$

For a discussion of the subject of notation, see the work of Gerischer [A 321]. With the above substitutions, equations (1-3a) and (1-3b) may be rewritten as follows:

$$i_c = nFk^+C_{M^+} \exp\left(\frac{-\alpha n\Delta\varphi}{\Phi}\right) \tag{1-4a}$$

and

$$i_a = nFk^-C_M \exp\left[(1 - \alpha)\frac{n\Delta\varphi}{\Phi}\right] \tag{1-4b}$$

A difficulty in the treatment of electrode reactions arises from the fact that one cannot measure electrical and chemical forces independently. Furthermore, $\Delta\varphi$ can be measured only relative to some reference state, *i.e.*, the voltage of a galvanic cell in connection with a reference electrode. The numerical values of k^+ and k^-, therefore, depend upon this electrode [A 321]. (As mentioned above, in dilute electrolyte, k^+ and k^- also involve a double-layer term.)

The use of a standard electrode of the same type as the electrode is physically significant. The concentrations of reactants at the "standard state" are indicated as follows (for simplicity, concentrations instead of activities are used):

$$^0C_M = {}^0C_{M^+} = {}^0C$$

The *standard exchange current* is thus defined as follows [A 321]:

$$^*i = nFk_1^+ \, {}^0C_M = nFk_1^- \, {}^0C_{M^+} \tag{1-5}$$

The advantage of defining $^*i$ in this manner is that it is independent of concentration and potential.

The potential at this standard state is E^0, *i.e.*, the *standard* redox potential, and E is the measured cell potential difference referred to an arbitrary reference electrode. To avoid introducing a constant correction term, the *numerical* value for E^0 is that referred to the same arbitrary reference electrode.

Incorporation of these definitions into equation (1-1) yields

$$i = {}^*i\left\{\frac{C_{M^+}(x = 0)}{{}^0C} \exp\left(-\alpha n\frac{E - E^0}{\Phi}\right)\right.$$

$$\left. - \frac{C_M(x = 0)}{{}^0C} \exp\left[(1 - \alpha)n\frac{E - E^0}{\Phi}\right]\right\} \tag{1-6}$$

where $C_{M^+}(x = 0)$ is the concentration of C_{M^+} at the surface of the electrode. In the absence of mass-transport problems (to be discussed later), C_{M^+} is the bulk concentration.

The following alternative development of an overall rate equation is taken from the work of Delahay [A 322]. The reference potential is chosen as the equilibrium potential of the electrode *operating reversibly* in the same solution. The overpotential η at current i is defined as follows:

$$\eta = E - E_{eq}$$

where E and E_{eq} are also measured with respect to the same reference electrode. The international (Stockholm) sign convention is used. Now consider equation (1-4a), in which we add $\alpha n(E_{eq} - E_{eq})$ to the numerator of the exponential term. Thus, with $\Delta\varphi = E$,

$$i_c = nFk^+C_{M^+} \left[\exp \frac{-\alpha n}{\Phi} (E_{eq})\right]\left[\exp \left(\frac{-\alpha n}{\Phi} \eta\right)\right] \qquad (1\text{-}7)$$

The equilibrium potential can be expressed by the Nernst equation as follows:

$$E_{eq} = E^0 + \frac{\Phi}{n} \ln \frac{C_{M^+}}{C_M} \qquad (1\text{-}8)$$

The insertion of equation (1-8) into equation (1-7) yields:

$$i_c = nFk^+(C_{M^+})^{1-\alpha}(C_M)^\alpha \left(\exp \frac{-\alpha n}{\Phi} E^0\right)\left(\exp \frac{-\alpha n}{\Phi} \eta\right) \qquad (1\text{-}9)$$

The "apparent exchange current" i_0 is that current which flows back and forth across the electrode–electrolyte interface although the net current i in the external circuit is zero. Thus,

$$i_0 = nFk^+(C_{M^+})^{1-\alpha}(C_M)^\alpha \exp \left(\frac{-\alpha n}{\Phi} E^0\right) \qquad (1\text{-}10)$$

It is possible to derive a comparison of i_0 and $^*i$ by including $(E_{eq} - E_{eq})$ in the exponential terms of equation (1-6). The following expression results:

$$i_0 = {^*i} \left(\frac{C_M}{{^0C}}\right)^\alpha\left(\frac{C_{M^+}}{{^0C}}\right)^{1-\alpha} \qquad (1\text{-}11)$$

These equations were derived for a simple first-order reaction. For more complicated, charge-transfer-controlled reactions, equation (1-7) takes the following general form:

$$i_c = nFk^+ \prod_i^n C_i{}^y \exp\left(\frac{-\alpha n E_{eq}}{\Phi}\right)\left(\exp\frac{-\alpha n}{\Phi}\,\eta\right) \qquad (1\text{-}12)$$

For example, consider the following reaction:

$$M^{+n} + ne^- + \delta X \rightleftharpoons MX_\delta$$

where X could represent a ligand, *e.g.*, NH_3 or H_2O.
Equation (1-4a) has the following form:

$$i_c = nFk^+(C_{M^+})(X)^\delta \left[\exp\left(\frac{-\alpha n}{\Phi}\right)\Delta\varphi\right] \qquad (1\text{-}13)$$

Proceeding as before, it can be shown that the apparent exchange current is given by

$$i_0 = nFk_1{}^+(C_{M^+})^{1-\alpha}(C_M)^{\delta(1-\alpha)}(C_{X_\delta})^\alpha \qquad (1\text{-}14)$$

The exponential term, $\exp(-\alpha n\, E^0/\Phi)$, is placed in $k_1{}^+$. Note that the concentration dependence of i_0 on C_{M^+} and C_X provides a means for evaluating α and δ.

Thus, in terms of the *apparent* exchange current i_0, equation (1-1) becomes

$$i = i_0 \left[\exp\left(\frac{-\alpha n F\eta}{RT}\right) - \exp\frac{(1-\alpha)n F\eta}{RT}\right] \qquad (1\text{-}15)$$

If i_0 is large, *i.e.*, the electrode reaction is "reversible," η is low (<20 mV) over a significant range of current and equation (1-15) can be linearized into the following form:

$$i = i_0\,\eta Fn/RT \qquad (1\text{-}16)$$

At elevated temperatures, the exchange currents are generally quite high. For example, at 450°C, in LiCl–KCl, the Zn/Zn^{+2} (1 M) couple has an exchange current of 150 A/cm². The system can be considered reversible even at high current drains and the reverse reaction must be considered, *i.e.*, equation (1-15) or (1-16) is used.

The same couple discharged in H_2O at 25°C has an i_0 of 2×10^{-5} A/cm². At high current drains (a few mA/cm²), the electrode reaction is effectively irreversible and the reverse action can be ignored, *i.e.*, the second exponential term can be discarded. The rate of this irreversible, activation-controlled process is given by the following equation:

$$i = i_0 \exp(-\alpha Fn\,\eta/RT) \qquad (1\text{-}17)$$

Equation (1-17) may also be written in the form

$$\eta = a + b \log i \qquad (1\text{-}18)$$

where

$$a = 2.303\,(RT/\alpha nF)\log i_0$$

and

$$b = -2.303\,RT/\alpha nF$$

This logarithmic relationship between the current i and the overvoltage η is known as the Tafel equation; i_0 and b are called Tafel constants.

Chemical Significance

Before proceeding to a discussion of diffusion effects, it is appropriate to consider the chemical significance of the terms in the electrochemical rate equations, such as equations (1-7) and (1-12). The terms that are characteristic of a particular chemical system are α, E_{eq}, and $k^{+,-}$, i.e., ΔG^*.

The activation energy is a measure of the stability of the activated electrolyte–electrode complex. Unfortunately, little direct information is available on the energetics of such species, except what may be inferred from the stability of the reaction products. The equilibrium potential E_{eq} is, of course, determined by the free energy difference between the starting material and the product of the charge-transfer step. The particular chemical state of the product ion is determined by the chemical composition of the electrolyte.

All anodic reactions involving a metal and an aqueous solution, in the absence of complex-forming or precipitating anions other than hydroxyl, follow one or another of the following general, overall processes distinguished by the fate of the metal cation [A 291]:

$$M \rightarrow M^{+n}_{(solv)} + n\,e^-$$

$$M + n\,H_2O \rightarrow M(OH)_n(s) + n\,H^+ + n\,e^-$$

$$M + n\,OH^- \rightarrow M(OH)_n(s) + n\,e^-$$

$$M + n\,H_2O \rightarrow MO_n^{-n}(aq.) + 2n\,H^+ + n\,e^-$$

$$M + n\,OH^- \rightarrow MO_n^{-n}(aq.) + n\,H^+ + n\,e^-$$

$M^{+n}(aq.)$ and $MO_n^{-n}(aq.)$ are to be regarded as typifying water-soluble cations and oxyanions, respectively, and as having appropriate hydrating water molecules associated with them. $M(OH)_n(s)$ typifies a sparingly solid hydroxide. The thermodynamic electrode potentials and solubility data for these reactions have been tabulated for specific metals (see, for example, the work of Pourbaix *et al.* [A 299]). Anions other than hydroxyl also tend to form, with metal cations, complexes of varying stability, or to produce sparingly soluble products. Many neutral molecules (*e.g.*, NH_3 and substituted amines) frequently form soluble complexes more stable than the aquo-complexes. The change in the thermodynamic equilibrium potential (E_{eq}) can, of course, be calculated from a knowledge of the equilibrium constant for the complex. Conversely, electrochemical potential measurements are used to evaluate complex stability and solubility product.

The fact that the final anodic reaction product is soluble does not mean that the reaction is necessarily sterically or kinetically easy, nor does a final solid product imply that it is necessarily sluggish. In general, those processes giving dissolved products progress more readily than those giving solids, since fluids generally do not block the anode effectively, whereas solids may often do so. Unless a high-surface-area structure is provided, the electrode reaction yielding a solid product will polarize rapidly, and the electrochemical coulombic efficiency will be low.

These thermodynamic influences of the electrolyte on the electrode discharge properties have been well-documented for aqueous systems; similar effects exist with nonaqueous electrolytes as well (see Chapter 4).

Electrode Reactions—Cathodic Processes. Electrolyte participation is also to be expected for cathodic processes. The discharge of most positive plates involves the eventual solvation of an anion. Water affords stability to these ions by hydrogen bonding and by the general electrostatic effect, such as occurs in ion–dipole interactions. This interaction between anions and water is very strong, particularly with anions such as fluoride and hydroxide, which are small and, therefore, have their negative charge concentrated in a small volume. The strength of this interaction is further increased, since the positive charge on the small hydrogen atom of water can come much closer to an anion than can a positive charge on an atom other than hydrogen.

In organic solvents, such as N, N'-dimethylformamide (DMF)

and dimethylsulfoxide (DMSO), the positive center of the solvent molecule is on an atom other than hydrogen, and the interaction is of the weaker ion–dipole type. The positive end of the DMF or DMSO dipole fits much less closely about the anion than does water and, as a result, anions (especially small ones) dissolved in DMF or DMSO have their energy decreased much less than when they are dissolved in water [A 300] (see Fig. 1-2). It is to be expected that the energy well for the discharged anion will be less deep in organic solvents, and possibly the activation energy will be increased as well. The cathodic discharge reactions are further complicated by the unfavorable physical properties of most of the materials used, *i.e.*, poor electrical conductivities, and significant solubilities.

Consider the following thermodynamic reaction cycle where Ag_2O is used as an example:

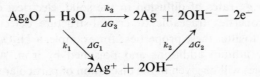

$$Ag_2O + H_2O \xrightarrow[\Delta G_3]{k_3} 2Ag + 2OH^- - 2e^-$$

$$k_1 \diagdown \Delta G_1 \qquad k_2 \diagup \Delta G_2$$

$$2Ag^+ + 2OH^-$$

These two routes to the product (silver) represent a method for a formal subdivision of the discharge mechanisms of positive electrodes.

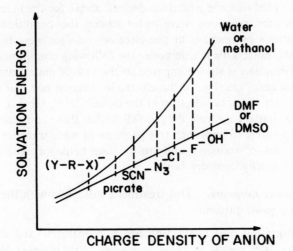

Fig. 1-2. Relative solvation energies of anions in water *versus* DMSO and DMF [A 300].

In the first $(1 + 2)$ path, the material discharges through a soluble intermediate, while in the second scheme the active material discharges directly. If reactions $1 + 2$ or 3 or both are reversible, then the electrode can be formally considered as an electrode of the "second kind," i.e., its potential is dependent on the anion concentration, in this case, OH^-.

The mechanism which functions in a given case is determined by the solubility of the material (ΔG_1) and the electronic conductivity of the oxides (or salts) involved. Mechanism 3 probably operates in the reduction of PbO_2, Hg_2Cl_2, and "Ni_2O_3;" mechanism $1 + 2$ seems to function, to some extent at least, in the reduction of insoluble salts, e.g., $PbSO_4$, $AgCl$, and $CuCl_2$, as well as the oxides of iron and cadmium [A 273].

The electrochemical behavior of materials discharging by mechanism 3 is determined, in large part, by their semiconductor properties, e.g., rates of diffusion in the solid, electrical conductivity, or other properties traceable to crystal-structure defects. Additives can be used to alter these properties; for example, a Ni_2O_3 electrode is activated by lithium additions and poisoned by iron. A number of other examples will be given in the discussion of particular depolarizers.

Mass-Transport Effects—Plane, Nonporous Electrodes

As pointed out, the equations derived above for the heterogeneous charge-transfer reactions were valid under the condition that the concentrations of reactant at the electrode surface were the same as those in the bulk. This is true under the following conditions: (1) if the electrode reaction is slow compared to the rate of mass transport, and (2) at time zero, i.e., before the electrode reaction has had an opportunity to alter the concentration at the double layer. Obviously, a large number of battery systems do not fall within these restrictions.

It is possible to approach this problem of mass transport from the point of view of transient and steady-state behavior. The discussion below will briefly consider both.

Transient Response. This treatment will begin with the following simplifying assumptions:

1. Migration and convection are not operative and diffusion is the only means of mass transport. This is generally true for systems with a large amount of inert electrolyte. Otherwise, migration must be considered.

2. Fick's laws of semi-infinite linear diffusion apply.
3. The reactant is diffusing from the bulk of solution to a *planar electrode*.
4. The electrode is operating at a constant current density.

These conditions correspond to the classical conditions of chrono-potentiometry; the derivations given below are based on treatments of this subject (see, for example, the work of Murray and Reilley [A 371]). The voltage–time plots derived correspond to curves of the type shown in Fig. 1-3(b). The derived equations must be considered as limiting cases of electrode performance since assumption 1 may not be rigorous after short times.

Consider a discharge reaction of the following type:

$$M^{+n} + n\,e^- \rightarrow M$$

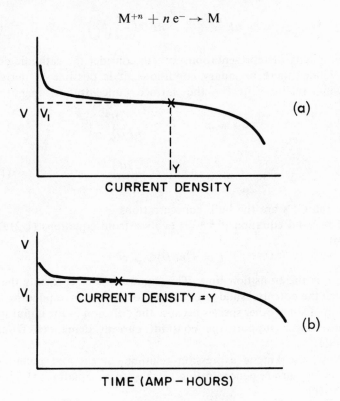

Fig. 1-3.   (a) General voltage–current curve at time 0.   (b) General voltage–time curve at constant current.

As current flows, M^{+n} is depleted in the vicinity of the electrodes and additional M^{+n} will diffuse in from the bulk of solution, *i.e.*, C_{M^+} will deviate from its equilibrium value. Eventually, a steady state will be achieved between the rate of the discharge reaction and the diffusion process.

The number of moles dN of substance that diffuse across a cross-sectional plane of unit area in the infinitesimal time interval dt is the flux. According to Fick's first law, the flux is proportional to the concentration gradient $\partial C / \partial X$ at the plane in question.

$$\frac{dN_{(x,t)}}{dt} = D\left(\frac{\partial C_{(x,t)}}{\partial x}\right) \tag{1-19}$$

and the current is given by the following relation:

$$i_c = nF\frac{dN_{(x,t)}}{dt} = nFD\left(\frac{\partial C_{M^+}}{\partial x}\right)_{x=0} \tag{1-20}$$

For simplicity of representation, we will consider the cathodic current only. Under these boundary conditions, it is possible to derive the following relationship for the surface concentration (*e.g.*, [A 323]):

$$C_{M^+(0,t)} = C_{M^+}^* - \frac{2it^{\frac{1}{2}}}{nFD_{M^+}^{\frac{1}{2}}\pi^{\frac{1}{2}}} \tag{1-21a}$$

and

$$C_{M(0,t)} = C_M^* + \frac{2it^{\frac{1}{2}}}{nFD_M^{\frac{1}{2}}\pi^{\frac{1}{2}}} \tag{1-21b}$$

where the C^*'s are the bulk concentrations.

The Sand equation [A 370,371] evolves from equations (1-21a) and (1-21b)

$$i = \pi^{\frac{1}{2}}nFD^{\frac{1}{2}}C_{M^+}\tau^{\frac{1}{2}}$$

where τ is the transition time. This characteristic represents the time at which the potential suddenly changes to a value corresponding to the reaction of some other species because the diffusion of the initial species can no longer support the constant current demanded from the electrode.

Now, the kinetic expression relating current and potential (in terms of the surface concentration C_{M^+}) is [from equations (1-7), (1-9), and (1-10)]

$$i = i_0\left\{\frac{C_{M^+}}{C_{M^+}^*}\exp\left(\frac{-\alpha n}{\Phi}\eta\right) - \frac{C_M}{C_M^*}\exp\left[\frac{(1-\alpha)n}{\Phi}\eta\right]\right\} \tag{1-22}$$

where

$$i_0 = nFk_3(C_{M^+})^{1-\alpha}(C_M)^\alpha$$

The dependence of the overpotential (at constant current) on time is obtained by substituting equations (1-21a) and (1-21b) into a linearized version of equation (1-15). Under conditions where $\eta \ll RT/nF$, the first two terms of the Maclaurin series can be substituted for each of exponential terms of equation (1-22), which yields [A 324]:

$$i = i_0 \left\{ \frac{C_{M^+}}{C_{M^+}^*} \left(1 - \alpha \frac{n}{\Phi} \eta \right) - \frac{C_M}{C_{M^+}^*} \left[1 + (1 - \alpha) \frac{n}{\Phi} \eta \right] \right\} \quad (1\text{-}23)$$

Substitution of equations (1-19) and (1-20) yields the following:

$$-\eta = \frac{\dfrac{i\Phi}{i_0 n} + \dfrac{2i\Phi}{nFn\pi^{\frac{1}{2}}} \left(\dfrac{1}{D_{M^+}^{\frac{1}{2}} C_{M^+}^*} + \dfrac{1}{D_M^{\frac{1}{2}} C_M^*} \right) t^{\frac{1}{2}}}{1 - \dfrac{2i}{nF\pi^{\frac{1}{2}}} \left(\dfrac{\alpha}{D_{M^+}^{\frac{1}{2}} C_{M^+}^*} - \dfrac{1-\alpha}{D_M^{\frac{1}{2}} C_M^*} \right) t^{\frac{1}{2}}} \quad (1\text{-}24)$$

By qualitatively considering equations (1-21a), (1-21b), and (1-22), it is clear that the overpotential should be exponentially related to $t^{\frac{1}{2}}$, since the surface concentration of C_{M^+} (directly proportional to $t^{\frac{1}{2}}$) is exponentially proportional to the overpotential. If C_M also varies with time, the curvature of η with $t^{\frac{1}{2}}$ is negligible, due to compensation between the linearized exponential terms. It is also apparent from inspection that equation (1-24) cannot be linearized into a term involving activation polarization ($i/i_0 n$) and a term involving concentration polarization.

Consider now the case where the reverse reaction is not significant, i.e., the process is irreversible. Equation (1-22) becomes

$$i = i_0 \left[\frac{C_{M^+}}{C_{M^+}^*} \exp \left(\frac{-\alpha n}{\Phi} \eta \right) \right] \quad (1\text{-}25)$$

and solving for η yields:

$$\eta = \left(\frac{-\Phi}{\alpha n} \ln i \right) + \frac{\Phi}{\alpha n} \ln i_0 - \frac{\Phi}{\alpha n} \left(\frac{C_{M^+}}{C_{M^+}^*} \right) \quad (1\text{-}26)$$

or, in the conventional Tafel form,

$$\eta = b \log i + a - \frac{\Phi}{\alpha n} \log \left(\frac{C_{M^+}}{C_{M^+}^*} \right) \quad (1\text{-}27)$$

The expression for $(C_{M^+}/C^*_{M^+})$ can be handled either in terms of the transient effects [equations (1-20), (1-21a), and (1-21b)] or in terms of a steady-state diffusion layer. Making the substitution from equation (1-21), one obtains

$$\eta = a + b \log i + \frac{\Phi}{2.3\alpha n} \log \left(1 - \frac{2it^{\frac{1}{2}}}{nFD^{\frac{1}{2}}\pi^{\frac{1}{2}}C^*_{M^+}}\right) \qquad (1\text{-}28)$$

Thus, the equation for the overpotential of an irreversible process is the linear combination of a kinetic term $(a + b \log i)$ and a concentration term of the form

$$\frac{\eta_c}{\alpha} = \frac{\Phi}{2.3n} \log \left(1 - \frac{2it^{\frac{1}{2}}}{nFD^{\frac{1}{2}}\pi^{\frac{1}{2}}C^*_{M^+}}\right)$$

so that

$$\eta = \eta_{\text{act}} + \frac{\eta_c}{\alpha} \qquad (1\text{-}29)$$

Note, however, that the concentration term still contains a kinetic parameter, *i.e.*, α [A 325].

Steady-State Behavior. An alternative representation of diffusion at a plane electrode is based on the idea that a steady-state diffusion layer will be formed at the electrode. This boundary layer is the zone of the liquid over which appreciable changes occur in concentration (diffusion boundary layer) or in flow velocity (hydrodynamic boundary layer). Without going into a detailed treatment of this subject, we can express the transport rate of a substance in terms of the effective thickness δ of the boundary layer. This effective thickness is a function of the flow velocity (free or forced convection), the kinematic viscosity, the electrode geometry, and the diffusion coefficient of the substance in question.

At the steady state,

$$i = nF \frac{D_R}{\delta} (C_{M^+} - C^*_{M^+}) \qquad (1\text{-}30)$$

for diffusion of C_{M^+} from the bulk to the surface where it is converted into product. D_R is the bulk diffusion constant for the diffusion of C_{M^+} through a diffusion layer of effective thickness δ. A similar equation applies to soluble products C_M.

The concept of a diffusion-limited current i_L represents an attempt to define the mass-transport characteristics of a given system in terms of a limiting condition, where the concentration of a reactant at the electrode surface approaches zero. Under these conditions, the diffusion-limited current is simply

$$i_{L,R} = nF \frac{D_R}{\delta} (C_{M^+}^*) \tag{1-31}$$

where $C_{M^+}^*$ is the bulk concentration. A large excess of indifferent electrolyte is assumed so that mass transport by migration may be ignored. At current densities less than the limiting one, the concentration at the electrode has a finite value, say C_{M^+}, which is expressed in terms of $C_{M^+}^*$ and i_L by the following relation:

$$\frac{C_{M^+}}{C_{M^+}^*} = (1 - i/i_{L,M^+}) \tag{1-32}$$

The deviation of the surface concentration from its bulk value will be reflected in the current–potential relation previously given. Introducing equation (1-32) into equation (1-31), we obtain

$$i = i_0 \left[\left(1 - \frac{i}{i_{L,M^+}} \right) \exp \frac{-\alpha F\eta}{RT} - \left(1 + \frac{i}{i_{L,M}} \right) \exp \frac{(1 - \alpha)F\eta}{RT} \right] \tag{1-33}$$

The Tafel constants derived from equation (1-33) are given in Table 1-I[A 16].

A problem with applying this concept is that many battery systems, particularly dry cells, do not have an "infinite" supply of electrolyte. Hence, the boundary layer is not constant, but continues to grow through the electrolyte space. In the case of an ionic product which diffuses away from the electrode, the limiting-current concept is also not useful; an upper limit for this process is not easily definable. Frequently, the concentration of a product can build up to several times its bulk concentration without much of a disturbing effect. At high enough concentrations, precipitation or dehydration occurs near the electrode, or the buildup of product may reduce the concentration of reactant. Of particular importance is precipitation which can lead to passivation in anodic reactions.

TABLE 1-I

Summary of Steady-State Limiting Cases for Redox Reactions at Solid Electrodes*

Mass transport limited	Equation	Limit	Plot	Slope	Intercept
No	$i = i_0 F\eta/RT$	$\eta < 10$ mV	i vs. η	$i_0 F/RT$	0
No	$\eta = [(2.3RT)/(\alpha F)][\log(i)/(i_0)]$	$\eta > 100$ mV	η vs. $\log i$	$\dfrac{2.3RT}{\alpha F}$	i_0
Partially	$i = [(1)/((1/i_0) + (2/i_L))][\eta F/RT]$	$\eta < 10$ mV; $\alpha \cong \frac{1}{2}$; $i_{L,M} = i_{L,M^+} = i_L$	i vs. η	$\dfrac{F/RT}{\dfrac{1}{i_0} + \dfrac{2}{i_L}}$	0
Partially	$\eta = [(2.3RT)/(\alpha F)][\log(ii_{L,M^+})/(i_{L,M^+} - i)(i_0)]$	$\eta > 100$ mV	η vs. $\log \dfrac{i i_{L,M^+}}{i_{L,M^+} - i}$	$\dfrac{2.3RT}{\alpha F}$	i_0
Completely	$\eta = [(2.3RT)/F] \times [\log(i_{L,M^+}/i_{L,M})(i_{L,M} + i)/(i_{L,M^+} - i)]$	—	η vs. $\log\left(\dfrac{i_{L,M^+}}{i_{L,M}} \times \dfrac{i_{L,M} + i}{i_{L,M^+} - i}\right)$	$\dfrac{2.3\kappa T}{F}$	0

* Taken from [A 16].

Many battery systems also involve the growth of solid reaction product on the surface, *e.g.*,

$$Cd + OH^- \rightleftharpoons Cd(OH)_2 + e^-$$

The rate equations must take into account this change of available surface area. The treatment of such a phenomenon is deferred until Chapter 7, which deals with battery charging, *i.e.*, electrocrystallization on a solid electrode.

Porous Electrodes

Highly porous electrodes or contained electrolytes present further problems in the analysis of reactant and product mass transport. In most analytical representations, convection is assumed absent and diffusion is the sole mode of transfer.

The treatment of this problem considers $\eta = f(x)$, where x is the distance along the pore. The first treatment given below is approximate [A 14], being restricted to situations where the electrode is effectively irreversible (low i_0 or high η). Mass-transport control and ohmic control are considered separately. A parameter ϕ is then derived which indicates the particular approximation that is applicable under a given set of circumstances. This treatment does have advantages in that the results are tractable and the method does illustrate the general considerations involved in treating porous electrodes.

For consistency with the original literature [A 14], R_b designates the reactant concentration within the bulk of the electrolyte, R_s is the concentration at the plane surface formed by the pore mouths, and R is the concentration within the pore (generally a variable).

The following simplifying assumptions are made:

1. The rate of the electrochemical process is determined by a charge-transfer step.
2. The electrode does not undergo change during the reaction. This, obviously, is only an approximation to a real battery plate.
3. The mass transport coefficients (D, resistivity, ion mobility) are constant despite changes in R.
4. The pores of the solid electrode are so interlinked that they behave as a uniform system.
5. The pores are small compared to the thickness of the electrode; consequently, gradients along the radius are negligible compared to gradients along the pore axis, *i.e.*, the system is one-dimensional.

6. The electrolyte completely penetrates the pores of the electrode.
7. The electrode is of uniform thickness, completely immersed in the electrolyte, and is bounded on one side by an electrically conducting, but otherwise inert, plate.
8. The bulk diffusion coefficient is pertinent.
9. The electrical conductivity of the electrode is high compared to that of the electrolyte.

Mass-Transport (Diffusion) Control. Consider an element of thickness Δx within a pore of length L. Since the element is at steady state, the amount of R reacting per unit time is equal to the amount of R entering the element per unit time at $x + \Delta x$ minus the amount of R leaving the element per unit time at x. Thus,

$$i_{\Delta x} = nF(R_{in} - R_{out})$$

Since

$$R_{in} = D_R \left(\frac{dR}{dx}\right)_{x+\Delta x}$$

and

$$R_{out} = D_R \left(\frac{dR}{dx}\right)_x$$

then

$$i_{\Delta x} = nF \left[D_R \left(\frac{dR}{dx}\right)_{x+\Delta x} - D_R \left(\frac{dR}{dx}\right)_x \right]$$

At this stage, we have assumed the pore diffusion coefficient to be equal to the bulk diffusion coefficient.

Now, from the definition of the derivative, the following equation results:

$$\frac{di}{dx} = nFD_R \left(\frac{d^2R}{dx^2}\right)$$

This equation is combined with a differential form of equation (1-6), i.e.,

$$di = i_0 s[(R/R_b) \exp(-\alpha\eta/b) - (P/P_b) \exp[(1 - \alpha)\eta/b]]dx$$

where b is equal to RT/nF and s is a surface roughness factor. It is assumed that the ohmic voltage gradient within the pore is small, and that η is of the same form throughout the pore. Thus, the solutions will apply only at low current densities. Upon integration under the proper boundary conditions, general expressions for i in terms of i_0, η, s, D, α, R, and P are obtained.

For small $i_0 s$ and for $\eta > 50$ mV so that the back reaction can be ignored, the equation reduces to the normal Tafel equation [equation (1-18)]. For large η or large $i_0 s$ or both, the equation reduces to one of the Tafel form, but with twice the slope:

$$i = (i_0 snFDR_b)^{1/2}(R_s/R_b) \exp(-\alpha\eta/2b) \tag{1-35}$$

For a negligible concentration gradient outside the electrode, *i.e.*, low currents, $R_s/R_b = 1$.

Thus, the modified apparent exchange current is

$$I_{0,M} = (i_0 snFDR_b)^{1/2} \tag{1-36}$$

and the apparent Tafel slope is twice the expected normal value.

Ohmic Control. In considering the ohmic effects alone, it is assumed *that R remains constant throughout the electrode* at a value R_s and that the ohmic potential gives rise to large changes in η. From Ohm's law,

$$i = \left(\frac{1}{\rho}\right) d\eta/dx \tag{1-37}$$

where ρ is the electrolyte resistivity (Ω-cm); from a differential form of equation (1-22) the following result is obtained:

$$\frac{d^2\eta}{dx^2} = \rho i_0 s \left\{\left(\frac{R_s}{R_b}\right) \exp(\alpha\eta/b) - \left(\frac{P_s}{P_b}\right) \exp[-(1-\alpha)\eta/b]\right\} \tag{1-38}$$

This equation is then integrated under the proper boundary conditions. Again, when i is small and η is large, the solution reduces to the normal Tafel equation:

$$i = (i_0 sF) \exp(\alpha\eta/b) \tag{1-39}$$

When $i_0 s$ is large, or when η is >50 mV, the following is obtained:

$$i = (2i_0 sb/\alpha\rho)^{1/2} \exp(\alpha\eta_s/2b) \tag{1-40}$$

The apparent exchange current is

$$I_{0,H} = (2i_0 sb/\alpha\rho)^{1/2} \tag{1-41}$$

and again the Tafel slope has double the normal value.

Combined Control. An important area of battery performance is that of high current drains, where it is necessary to consider mass transport and ohmic effects operating simultaneously. The two effects are combined by equating the derivative of Ohm's law [equation (1-37)] with the mass transport [equation (1-34)] and solving for R/R_s, where R is the concentration within the pore:

$$\left(\frac{1}{\rho}\right) d\eta/dx = nFD \int_0^x (d^2R/dx^2)\, dx \qquad (1\text{-}42)$$

It is assumed that $\int d\eta = \rho FD \int dR$. In other words, there is a linear relationship between η and R, which, of course, is only an approximation [see, *e.g.*, equations (1-23) and (1-28)].

$$R/R_s = 1 - (\eta_s - \eta)/R_s\rho nFD \qquad (1\text{-}42a)$$

This result is inserted into equation (1-21) and integrated under the condition that

$$\left(\frac{P_s}{P_b}\right) \exp[-(1-\alpha)\eta/b]$$

is negligible, *i.e.*, the electrochemical process is irreversible.

At this point, an auxiliary parameter is defined:

$$\phi = R_b\rho nFD_s(\alpha/b) \qquad (1\text{-}43)$$

From the solution of the integral, it can be shown that, depending on the value of ϕ, the exchange current will be given by either equation (1-36) or (1-41).

The mathematical expression for the intermediate case is rather complex. For values of $\phi \geqslant 5$, the exchange current $I_{0,T}$ is given by $I_{0,H}$; for values of $\phi \leqslant 0.5$, the exchange current is given by $I_{0,M}$. The proportionality between $I_{0,T}$ and $I_{0,H}$ for intermediate ϕ is given below:

ϕ	$I_{0,T}/I_{0,M}$
>5	1
5	0.9
2	0.75
0.5	0.46

Summary. Thus, $\phi = \rho nFDR_b\alpha/b$ can be calculated from known properties of the electrolyte and from measured properties of the plane electrode (α/b). The value of ϕ determines the apparent value of the exchange current, *i.e.*,

$$I_{0,M} = (i_0 snFDR_b)^{1/2} \qquad \phi < 0.5 \qquad (1\text{-}36)$$

$$I_{0,H} = (2i_0 sb/\alpha\rho)^{1/2} \qquad \phi > 5 \qquad (1\text{-}41)$$

In both cases, the Tafel slope is double the normal slope.

Example. Consider the discharge of an hydroxyl ion at a nickel oxide anode. A "double" Tafel slope of 0.065 V/decade was obtained. The value of α/b is 13.5, and since the electrolyte used was 30% KOH, $\rho \cong 2\ \Omega\text{-cm}$, $D \cong 5 \times 10^{-5}\ cm^2/sec$, and $R_b \cong 5 \times 10^{-3}\ mole/cm^3$.

$$\phi = \rho R_b nFD\alpha/b$$
$$\phi = 2(5 \times 10^{-3})(1 \times 10^{-5})(5 \times 10^{-5})(13.5) = 0.7.$$

Thus, the reaction involves both ohmic and mass-transport control in the current-density region where the Tafel slope is 0.065.

Current Distribution. At any given time, there can be a range of reaction rates within the pores. This distribution of rates will depend upon physical structure, conductivity of the matrix and electrolyte, and parameters characterizing overpotential phenomena. In developing equations to describe this effect [A 15], the same pore model is used, *i.e.*, a one-dimensional pore, backed by a solid metallic conductor of negligible resistivity. Again, *one-dimensional* describes quantities, such as overpotential, current density, and concentration, that vary with depth within the electrode but not with lateral position.

For simplicity of calculation, it is desirable to approach the problem separately for ohmic control and for mass-transport control. Again, consider the case where R remains constant through the pore and is equal to R_b, *i.e.*, the system is under ohmic control. (Assumption 9, *i.e.*, low electrode resistance, is abandoned.) The basic kinetic equation is used in the following form [A 15]:

$$i = i_0 \exp\left[\frac{\alpha}{b}(\phi_1 - \phi_2)\right] \qquad (1\text{-}44)$$

where ϕ_1 is the electrostatic potential of the electrode and ϕ_2 is the electrostatic potential of the solution. Although these potentials are not measurable, changes in potential, $\Delta(\phi_1 - \phi_2)$, are accessible. Note also that the reverse reaction has been considered negligible, *i.e.*, Tafel kinetics.

The derivative of equation (1-36) is combined with Ohm's law [equation (1-37)]. The following relationships, based on Ohm's law and electrical neutrality, are used to eliminate i_2, ϕ_1, and ϕ_2:

$$i_2 = -\chi \frac{d\phi_2}{dx} \qquad i_1 = -\sigma \frac{d\phi_1}{dx} \qquad \frac{di_1}{dx} + \frac{di_2}{dx} = 0$$

where χ is the effective conductivity of the pore electrolyte; σ is the conductivity of the electrode material; i_1 is the current in the matrix phase; and i_2 is that in the pore electrolyte. The resulting equation is:

$$\frac{d^2 i_1}{dx^2} = \frac{di_1}{dx} \beta \left[\frac{I}{\chi} - i_1 \left(\frac{1}{\chi} + \frac{1}{\sigma} \right) \right] \qquad (1\text{-}45)$$

The following boundary conditions are used:

$$i_2 = I, \, i_1 = 0, \, \phi_2 = 0 \qquad \text{at } x = 0$$

$$i_2 = 0, \, i_1 = I \qquad \qquad \text{at } x = L$$

and, by definition,

$$\beta = (1 - \alpha) \frac{nF}{RT}$$

The free solution–porous electrode interface is taken *versus* $x = 0$.

These conditions state that at the electrode–solution interface the current I is carried entirely by the pore electrolyte, while at the backing the current is carried entirely by the matrix. As an arbitrary reference of potential, $\phi_2 = 0$ at $x = 0$.

Upon introduction of the quantities

$$y = \frac{x}{L}$$

$$j = \frac{i_1}{I}$$

$$\delta = L \, | \, I \, | \, \beta \left(\frac{1}{\chi} + \frac{1}{\sigma} \right)$$

$$\epsilon' = \frac{L \, | \, I \, | \, \beta}{\chi}$$

the following dimensionless equation is obtained:

$$\frac{d^2 j}{dy^2} = \frac{dj}{dy} (\delta j - \epsilon') \qquad (1\text{-}46)$$

Under the boundary conditions $j = 0$ at $y = 0$ and $j = 1$ at $y = 1$, the dimensionless reaction rate is found to be

$$\frac{dj}{dy} = \frac{I}{L}\frac{di}{dx} = \frac{2\theta^2}{\delta}\sec^2(\theta y - \psi) \qquad (1\text{-}47)$$

where θ is an integration constant.

The item of interest is the variation of dj/dy with y (in accord with the boundary conditions $j = 0$ at $y = 0$). A plot of equation (1-47) is given in Fig. 1-4 for the situation where $\epsilon' = 0$, i.e., for an electrolyte

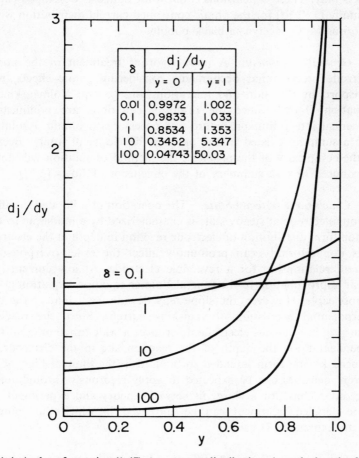

Fig. 1-4. A plot of equation (1-47), i.e., current distribution along the length of the pore under ohmic control [A 15].

with a high conductivity χ. At low currents, the reaction gradient is uniform across the length of the pore, *i.e.*, the same amount of charge is generated in each element dy. However at high currents the reaction takes place primarily at the pore–metal backing plate interface. In the particular situation where the conductivities of the pore and the electrolyte are equal, it is found that the reaction takes place predominantly at both interfaces. With a low electrolyte conductivity, the reaction takes place primarily at the pore–solution interface. As mentioned, these conclusions were reached by considering ohmic effects only. These conclusions confirm the concepts developed experimentally [A 269,270] for the ohmic-controlled current distribution within a porous MnO_2–acetylene black bobbin.

General Treatment. A more rigorous treatment of the porous electrode problem has been made, considering ohmic effects, mass transport by diffusion and migration, and general Volmer kinetics [equation (1-16)]. Since the resulting equations are nonlinear, a numerical integration procedure was necessary to obtain a solution. The assumptions listed in the previous development apply. Because of the complexity of the problem, the method of solution will not be reproduced here; a summary of the conclusions follows [A 271].

Conclusions—Steady State. The operation of a flooded, nonflow, porous electrode at steady state is characterized by a moderate to high nonuniform distribution of electrode reaction in depth in the electrode. This nonuniformity can profoundly affect the exact overpotential–current relationship for a reversible electrode. At low currents, the linear approximation to the general Volmer equation [equation (1-16)] is applicable. However, the slope may deviate from that of the local overpotential expression at vanishing current. Since the reactions occurring in a porous electrode do so over a wide range of conditions dependent upon the depth of the reaction site in the electrode, no reduced kinetic form intended to be used in the limit of high or low current densities can be expected to apply rigorously throughout the electrode. Thus, for a closer fit between theory and experiment, this more detailed treatment can be employed in which the complete kinetic expression is used.

Conclusions—Transient Effects. A second set of conclusions resulting from this treatment concerns the transient behavior of the porous electrode. The transient behavior of a porous electrode system is taken to be the behavior over the period from completion of the circuit until steady state is achieved at constant current drain.

There are effectively two time-dependent phenomena, one involving the charge–discharge of the electrical double layer and the second involving the establishment of a steady state for the mass transport of material.

The activation transient, *i.e.*, the time required for the double layer to come to equilibrium, is illustrated by the following calculation. In the vicinity of equilibrium, the overvoltage–current relationship is

$$\eta = \left(\frac{RT}{nFi_0}\right) i$$

where η is the overvoltage, i_0 is the exchange current density, i is the current density, and the other symbols have their usual meaning. As an example, consider an electrode reaction at 25°C with $i_0 = 10^{-10}$ A/cm², $n = 4$ (*e.g.*, O_2/H_2O in acid), and $\eta = 1$ and 50 mV at the initial and steady states, respectively. The time constant for discharging the double layer through an effective resistance (RT/nFi_0) is

$$\tau = \left(\frac{RT}{nFi_0}\right) C_{dl} \qquad (1\text{-}48)$$

and the increase in overpotential is given by

$$\eta = \eta_0 e^{-t/\tau} \qquad (1\text{-}49)$$

If it is assumed that $C_{dl} = 100 \ \mu\text{F/cm}^2$ for a rough surface, then $\tau = 6 \times 10^7 \times 100 \times 10^{-6}$ or 6×10^3 sec. Thus, for example, the time required for the overpotential to increase from 1 mV to a steady-state value of 50 mV is approximately 2 hr. If the exchange current is of the order of 10^{-3} A/cm², a similar calculation indicates approximately 10^{-4} sec for equilibration of the double layer.

Calculations of the mass-transport transients indicate that these processes are characterized by times of the order of $10–10^4$ sec to, say, 90% of the total overpotential change. "Since many applications of porous electrodes involve operating periods which are not long compared to this time range, mass-transport transient behavior should be carefully considered in analysis of cells involving porous electrodes. Operation in the steady state could be the exception rather than the rule"[A 271].

The magnitude of the characteristic time is relatively constant over a wide range of operating conditions for cases where migration is the most significant reactant transport mechanism, *i.e.*, for binary electrolytes. In this case, changes in concentration have no effect on the dimensionless representation, and increasing current increases reactant transport as well as consumption. For cases where migration is only a secondary means of reactant transport, excess inert electrolyte being present, the durations of the transient phenomena are influenced by the manner of operation. Kinetic parameters have a relatively small influence on the characteristic times. Decreasing the depth of the electrode reduces the duration of transients. In those deep portions of the electrode where current densities are small, the reactant content is but slowly depleted. However, reactant supply from the electrolyte in the depth of the electrode to the areas near the face, where reaction rates are high, is usually quite small compared to supply from the electrode face. Thus, the presence of a reservoir of reactant in the pores significantly affects the course of the process only in those parts of the electrode where very little current is transferred.

Analytical Representations of Battery Discharge

Attempts have been made to describe the time dependence of a complete battery discharge by empirical equations. Perhaps the most well-known is the Peukert equation:

$$I^n t = C$$

where n and C are empirical constants, functions of the particular cell materials and cell construction. Although the equation does summarize performance over a range of discharge conditions, the constants have no theoretical basis and must be individually determined.

A more recent attempt at this problem was set up as follows [A 274]. First, it was assumed that the overpotential and current are linearly related:

$$\eta = \frac{RT}{nFi_0} i$$

This is fairly accurate up to values of η equal to 0.03 V. Apparently, the variation of η with i can be fitted by a straight line within approximately 0.02–0.04 V up to values of η equal to 0.2–0.4 V. A potential drop of this magnitude would cover the major portion of the polarization that occurs during most battery discharges. This equation may be rewritten as follows:

$$E_{\text{cell}} = E^0 - Ki_a \qquad (1\text{-}50)$$

where i_a is the current per square centimeter of active material remaining at time t.

In the case of a porous electrode, i_a was assumed to be inversely proportional to the amount of unused active material and also to be equal to i at the beginning of the discharge. Thus,

$$i_a = \left(\frac{Q_c}{Q_c - it} \right) i \qquad (1\text{-}51)$$

where t is the time at any point during the discharge, Q_c is the amount of active material initially available, and i is the current per square centimeter of total plate (used and unused material). The potential drop due to internal resistances is defined as Ni, where N is the internal resistance per unit area. Consequently,

$$E_{\text{cell}} = E^0 - K \left(\frac{Q}{Q - it} \right) i - Ni \qquad (1\text{-}52)$$

The equation in this form has been used to fit the discharge curves for many batteries over a range of current densities. However, in applying this equation, it was necessary to consider the terms E^0, K, Q, and N as adjustable constants. The E^0 term agreed reasonably well with the open circuit potential (generally ± 0.05 V), and apparently the Q term does approximate the capacity of the plates. However, the physical meaning of the computed K and N is somewhat obscure; e.g., some negative values have been obtained for N. Thus, the equation in its present form must be regarded as empirical in nature, with at least two adjustable constants.

REFERENCE ELECTRODES

The experimental evaluation of individual electrode characteristics is, of course, based on the availability of a stable, known reference point. The availability of such an electrode is no longer a problem in aqueous electrolytes. However, organic electrolytes are another matter.

In most battery-oriented applications, the reference electrode is used (1) to measure the Gibbs free energy of a reaction, *i.e.*, to establish an order of potentials in a given medium, and (2) to establish the changes in potential of a working electrode.

The subject of reference electrodes has been treated elsewhere in considerable detail [A 158]. The present discussion is intended to emphasize the necessity of using valid reference electrodes in evaluating the performance of active battery material.

A reference electrode is a "hypothetical, completely reversible, nonpolarizable electrode, the potential of which is unaffected when electric current flows across the electrode–solution interface. It is desirable that the electrode reaction be simple and well-defined. Under these constraints, the reference electrode, of course, will obey the Nernst equation over a wide range of concentration."

"If the electrode is not reversible, a number of problems can arise. For example, (1) the electrode potential will not be reproducible from cell to cell; and (2) since the potential is not controlled by the principal ion in solution, it is to be expected that the potential will become unstable and indefinite, not so much controlled by, as at the mercy of, dissolved oxygen and such other impurities as the solution may contain."

"In real systems processes of chemical change and transport of matter must keep pace with the flow of electricity and do not occur in practice either unhindered or in the absence of a gradient of chemical potential." The acceptable degree of approximation to nonpolarizable is, of course, determined by the necessary accuracy of the measurement.

"In the vicinity of equilibrium, the current is given by

$$i = i_0 \eta n F / RT$$

or

$$\eta = \frac{RT}{nF} \frac{i}{i_0}$$

and where i_0 is the exchange current and the other symbols have their usual meaning. The practical condition for a real electrode to be thermodynamically well-behaved is that its exchange current should be large compared to any net current that it is required to pass in use. If, in a particular case, the exchange current should fall too low (*e.g.*, by poisoning the electrode surface), not only may the electrode become polarizable, but also it may no longer be able to attain its proper equilibrium potential."

"Exchange current densities for various kinds of metal–solution interfaces, at room temperature, cover a range of about 10^{-2} to 10^{-8} A/cm², but the useful range for reference electrodes is normally much more restricted than this; it will depend in part on the sensitivity of the measuring instrument to be used" [A 158].

If the exchange current is of the order of 10^{-3} A/cm², it can be shown that the overpotential would decay from 10 to 0.1 mV in 10^{-4} sec. To avoid polarization exceeding 0.1 mV, a current density of the order of 10^{-5} A/cm² should be used in the measurement.

"The measurement of exchange current is neither easy nor feasible as a normal ancillary to potentiometric work, except in one rather *ad hoc* way which may not always be easy to apply. In the close vicinity of the equilibrium potential, the net current passed by an electrode forms a linear plot against the displacement of the potential from the equilibrium value and the slope of this plot is proportional to the exchange current. Clearly, to determine such a slope (or, rather, its inverse) is no more than a common-sense way of determining the polarizability of an electrode. Provided that the other electrode of the cell is known to be less polarizable, this can be done in a purely comparative way in terms of potentiometer settings and galvanometer deflections on either side of the null point. Any hysteresis effects which are observed indicate gross irreversibility. Under suitable circumstances, this procedure can give quite useful information" [A 158].

ELECTROLYTE EFFECTS

The specific roles which the electrolytes must fulfill in the charge and discharge of a battery can be quite complex. For example, the solvent in a battery system is more than just the required electrolyte path for the electrochemical reaction. It is a medium for the transfer of material (ionic and unchanged species) through the bulk of solution into the electrical double layer and for transport of electrode reaction products away from the electrode. Often it participates directly in the half-cell reaction, *i.e.*, molecules of the solvent are involved in the transition complex which controls the reaction kinetics. Finally, the solvent may be involved in unwanted side reactions (*e.g.*, dissolution) which shorten the life of the battery.

Studies of these processes have recently been extended from the conventional aqueous battery electrolytes to a wide variety of nonaqueous solutions. As will be indicated in Chapters 3 and 4, this is due to the intense interest in aprotic solvents which are thermodynamically compatible with highly active negative materials, such as magnesium and lithium. The electrolytes being studied encompass a wide range of physical properties, such as dielectric constant, viscosity, and density. Quantitative description of ionic conductivity in terms of these parameters has proved to be a difficult problem. General, background information for this area is presented below.

Studies of ionic conductivity classically have sought to answer the following two questions: What is the structure and composition of the ion in solution? What is the physical mechanism by which ionic conduction takes place?

Ion Composition and Structure

The effective size of an ion in solution (*i.e.*, the entity which migrates under the influence of an external electric field) is, to a first approximation, determined by the electrostatic solute–solvent and solute–solute interactions.

In dilute solution, the ions are well-separated by electrolyte. An average separation of the ions can be estimated by assuming, as a rough guide, that the ions in solution are arranged in a cubic lattice, at least as a time average. For a 1:1 electrolyte at a concentration of c moles/liter, the average interatomic distance is $9.40 \, c^{1/3}$ Å, giving rise to the results shown in Table 1-II [A 150].

TABLE 1-II

Average Separation of Ions in a Solution of a 1:1 Electrolyte*

c (moles/liter)	0.001	0.01	0.1	1.0	10.0
Separation (Å)	94	44	20	9.4	4.4

* Taken from [A 150].

If the dielectric constant of the solvent is sufficiently high, the ions in dilute solution are electrostatically isolated from one another and behave in an independent fashion. The structure of the ion in this dilute solution is thus determined almost solely by the *interactions with the solvent*.

The *electrostatic* interactions are usually discussed in terms of the dielectric constant. (The more specific chemical interactions, such as hydrogen bonding and direct coordination, will be treated in a later section.) Consider two point charges in a vacuum separated by a distance r. The potential energy between the charges is found to be (Coulomb's law):

$$qq'/r$$

If a dielectric medium is placed between the two charges, the potential energy between these two charges is now less, *i.e.*,

$$qq'/Dr$$

where D is the dielectric constant. The difference in potential energy because of the dielectric is

$$\frac{qq'}{r}\left(1 - \frac{1}{D}\right) \tag{1-53}$$

The question is: What molecular processes within the dielectric have absorbed this energy?

Much of the literature discusses this subject in terms of diminishing (polarizing) the electric field between charges, rather than in absorbing energy from the field. The two points of view are, of course, equivalent. Polarization is here defined as a process by which energy can be absorbed from the field that is induced by the existence of charged ions. In effect, these processes can be regarded as energy sinks. The more processes available, the more energy withdrawn from the field and the higher is the dielectric constant.

The first polarization generally considered is that due to the distortion of electronic distributions within molecules and is referred to as "molecular polarizability." A second type of polarization is that due to the partial lining up of already existing permanent molecular dipoles, a process called "orientation polarization." The rotation of these dipoles can be hindered by short-range chemical interactions, such as hydrogen bonding. Hence, a larger amount of energy is abstracted from the field to counteract these interactions and produce the orientated dipoles.

It is possible to estimate the degree of these short-range solvent association effects from measurements of the dielectric constant, index of refraction, and dipole moment. This estimate is generally made in terms of Kirkwood's g factor [A 150]. Acetone and chloroform, believed to be unassociated liquids, give $g = 1.0$; the strongly associated liquid hydrogen cyanide gives $g = 3.6$; water gives $g = 2.6$; nitrobenzene and pyridine give $g = 0.8$ and 0.7, respectively, indicating "contra-association" of the dipoles as opposed to head-to-tail or "co-association" in hydrogen cyanide.

The dielectric constant is a macroscopic property of matter—a measure of the ability of all solvent particles to orientate uniformly within a field. To some extent, the dielectric constant is a function of field strength. It is obvious that the processes described will absorb only a finite amount of energy, after which the dielectric will be "saturated." The result will be a decrease in the effective D.

Consider the situation of charged ionic species in a dielectric medium, such as water. At the small distances involved, the field intensity is of the order of 10^6V/cm; Coulomb's law, even if we insert the bulk dielectric constant of water ($\sim$80), gives a field of 0.5×10^6 V/cm at a distance of 6 Å from the center of a univalent ion. This field is sufficiently intense to cause a marked dielectric saturation in surrounding water molecules. Thus, in the first layer of molecules around an ion, the bulk dielectric constant of the medium is not relevant. The molecules in the second layer ($\sim$5 Å out) will be less strongly orientated, and the effective dielectric constant rises. There is a consequent reduction in the total average dielectric constant as measured by an external applied field. As a result, the dielectric constant of an electrolyte solution falls as the concentration, and, hence, total ion–solvent interaction is increased [A 150].

It is found for many electrolytes (1:1, 2:1, 1:2, 3:1 valency types) that the dielectric constant falls linearly as the electrolyte concentration is increased. This linear drop holds in most cases up to about 2N, after which, in the case of NaCl at least, the drop is less than linear. The general equation developed for this situation is

$$D = D_w + 2\delta c \tag{1-54}$$

where δ is a function of the solute. For NaOH, $\delta = -10.5$, so that for 2N sodium hydroxide

$$D = 78.3 - (2)(10.5)(2) = 36$$

Ion Association. In circumstances where the energy of mutual electrical attraction of ions is substantially greater than their thermal energy, *e.g.*, at low dielectric constant or high solute concentration, the ions can form entities in solution of sufficient stability to persist through a number of collisions with solvent molecules. In the case of a symmetrical electrolyte, such ion pairs will have no net charge; in effect, the solute is undissociated. In the case of unsymmetrical electrolytes, the new ion pair will be charged and will contribute to the conductivity, though less than would its constituent ions in a free state.

One estimate of the critical distance for ion-pair formation is given by the following relation:

$$r_c = \frac{z_1 z_2 e^2}{2DkT} \qquad (1\text{-}55)$$

At this distance, the mutual electrical potential energy of the two ions is equal to $2kT$. Some values for r_c are given in Table 1-III [A 150].

TABLE 1-III

Critical Distance of Approach for Ion-Pair Formation

Solvent	D	$r_c(\text{Å})$
H_2O	78.3	3.52
Acetonitrile	36.7	7.45
Pyridine	12.0	23
Benzene	2.27	121

Thus, the dielectric constant is a measure of the ability of the solvent to shield the individual ions from one another. In an overall sense, the interactions involved can be considered electrostatic, in that they are independent of the specific ions present.

"The physical picture which has been evolved is as follows: About each ion is a sheath of solvent oriented by the electrical field of the center ion. In the case of noble-gas type ions and for any water molecules not in immediate contact with the ion, this hydration (or solvation) may be far from permanent in the everyday sense of the word; rather, the permanence implied is relative to the time scale of the Brownian motion. It is to this ion–solvent sheath that the Debye–Hückel theory applies" [A 150].

Chemical Interactions. There are a few cases where these electrostatic considerations are insufficient to explain the observed behavior of ions. Apparently, in the case of aqueous solutions of chromium and cobalt ions, the inner sheath of water molecules is firmly attached, possibly by coordinate links. A recent paper [A 349] explains solute conductivity differences in dimethoxyethane and tetrahydrofuran in terms of the ionic coordination with the solvent. Chemical interactions will most likely have to be invoked to explain the behavior of certain mixed organic solvents. One such system is shown in Table 1-IV. Another involves the interaction of propylene carbonate

TABLE 1-IV

Conductivities of 0.63 M LiAlCl₄ Solutions*

Solvent	Conductivity $(\Omega\text{-cm})^{-1}$	$D^{\dagger}$	$\eta^{\dagger}$
PC	6.6×10^{-3}	64	2.5
PC + EE‡ (1:1)	9.8×10^{-3}		
EE	2.7×10^{-3}	4.3	0.23
EE (2.5 M)	12.2×10^{-3}		

* Taken from [A 252].
† For the pure solvent.
‡ Ethyl ether.

(PC) and ethylene carbonate (EC). It has been reported [A 249] that mixtures of EC and PC give conductivities higher than those that can be obtained with either solvent alone. For example, KPF_6 in 80% EC–20% PC solvent had a conductivity of 1.16×10^{-2} $(\Omega\text{-cm})^{-1}$. This increase is accompanied by a decrease in viscosity and a 35% increase in dielectric constant.

Viscosity. The effective size of an ion is determined by the dielectric constant and specific solvent–solute interactions. The movement of this ion aggregate through solution, *i.e.*, ionic conductivity, is affected by the viscosity of the medium. In dilute solution, the viscosity is presumably determined by solvent–solvent interactions. However, in more concentrated solutions, viscosity is also influenced by the ion–solvent and ion–ion interactions.

Viscosity is the force required to produce unit rate of shear between two layers separated by unit distance. For a particle of macroscopic dimensions moving in an ideal hydrodynamic continuum, it is correct to calculate the frictional resistance in terms of the dimensions of the particle and the macroscopic viscosity of the medium, *i.e.*, Stokes' law applies:

$$v = F/6\pi\eta r \tag{1-56}$$

where η is the viscosity, F is the Faraday's constant, and v is the ionic velocity. Solving for r_{λ^0} (Stokes' radius), with v in terms of the limiting equivalent conductivity, we obtain:

$$r = |z| F^2/(6\pi N\eta^0\lambda^0) \tag{1-57}$$

The applicability of Stokes' law to ionic movement requires the ion to be large with respect to the components of the medium through which it is moving. Indeed, there is reason to believe, from an examination of the temperature dependence of ionic mobility, that for ions which are (1) intrinsically large and of low surface charge, or (2) of sufficiently large surface charge to form firmly hydrated entities, Stokes' law is of the correct form, though the numerical constant may not be 6π. For these ions, the product $\eta^0\lambda^0$ is relatively constant over a fair range of temperatures in water. From a study of tetraalkyl ammonium ions, it appears that Stokes' law is applicable in water for particles greater than 5 Å in radius. Included in this list are polyatomic ions, such as acetate, substituted ammonium ions, and extensively hydrated ions, such as Li^+, Ca^{+2}, and La^{+3}. The monatomic ions K^+, Rb^+, Cl^-, Br^-, and I^-, and ClO_4^- and NO_3^- show some deviation. Nevertheless, the ordinary viscous forces still account for most of the resistance to the motion of these ions in water, though there is evidently some other effect operative as well [A 150].

Walden's rule states that $\eta^0\lambda^0$ is constant for a given solute. It is seen from the above equation for Stokes' radius that

$$\eta^0\lambda^0 = \frac{1}{r}\left(\frac{|z| F^2}{6\pi N}\right) \tag{1-58}$$

Thus, for $\eta^0\lambda^0$ to be a constant, the ionic radius must also be a constant and Stokes' law must apply. As shown above, this would be so in the absence of solvation. In a given solvent, constancy with temperature would be observed for solvated ions, if the solvent sheath were large and tightly bound.

A specific question is how the mobility of ions is related to the change in viscosity of the solution, bearing in mind that this changed viscosity is itself produced by the ions of interest. This change in viscosity is usually expressed in terms of the relative viscosity:

$$\eta_{rel} = \eta/\eta^0 \tag{1-59}$$

where η^0 is the viscosity of the pure solvent at the same temperature. Unfortunately, the same symbol is used in the literature for viscosity and overvoltage. This custom will also be employed here, since it is felt that little confusion will actually result:

$$\eta_{rel} = 1 + A_1(C)^{1/2} + A_2 C \tag{1-60}$$

The constant A_1 is a function of solvent properties, ionic charges and mobilities, and temperature. Numerically, A_1 is fairly small, e.g., 0.005 for KI in water at 25°C. The A_2 coefficient is highly specific for the electrolyte and temperature. The absolute value of A_2 is somewhat larger, e.g., −0.014 for KCl at 25°C.

These A_2 coefficients are additive properties of the constituent ions and are strongly correlated with the entropy of solution of the ions. Negative values are found with those ions which exert a "structure-breaking" effect on water, e.g., Rb^+, Cs^+, and NO_3^-. These negative values appear to be confined to aqueous solutions, and even here they seldom cause a decrease of more than 10% in the viscosity. More typical are fairly large positive values of A_2 found with ions which are strongly hydrated; e.g., at 25°C, the A_2 values for Na^+, Li^+, Mg^{+2}, and La^{+3} are 0.0863, 0.1495, 0.3852, 0.5888, respectively.

Quantitative Description of Conductivity

Onsager Equation. The classical theoretical equations for ionic conductance are attempts to express the conductivity of an ion (or, more specifically, the deviations from conductivity at infinite dilution) in terms of properties of the medium (dielectric constant and viscosity) and properties of the ions (charge, radius, and concentration).

One of the more well-known of these is the Onsager equation; the basic reasoning involved is as follows. In dilute solutions, interionic attractions and repulsions lead to two effects—the "electrophoretic effect" and the "relaxation effect"—both of which result in the lowering of ionic mobilities with increasing ion concentrations.

Consider, first, the electrophoretic effect. An ion, or more accurately a solvated ion, is in turn surrounded by an "ionic atmosphere" distributed with radial symmetry around the ion as center. This ion atmosphere is due to the fact that interionic attractions and repulsions tend to produce a slight preponderance of negative ions in the vicinity of a positive ion and *vice versa*. Although the ion atmosphere is treated as a reality in mathematical discussions, it actually is the result of a time average of a distribution of the ions. Each ion serves as a center of an ion atmosphere, and the relative position of each ion with respect to the other charged bodies in the solution influences the atmosphere of all other ions.

Under the influence of an electric field, the center ion will tend to move in a given direction. However, the ion atmosphere, being oppositely charged to the ion itself, will tend to move in a reverse direction, constituting a drag on the ion under consideration. In the mathematical derivation, it is assumed that this imposed electric force acting on the ion atmosphere will produce a motion of the solvent. Thus, the fundamental explanation of this electrophoretic effect must be sought in a modification by the ion atmosphere of interactions between ions and solvent.

The relaxation effect can be explained as follows. Around a selected ion, there is an ionic atmosphere of spherical symmetry. If the ion is suddenly moved, the atmosphere behind the ion will tend to collapse, while a new atmosphere will tend to build. However, this collapse is not instantaneous and will exert an electrostatic attraction in a contrary direction to the motion of the ion.

The Onsager equation, derived from these considerations, can be written in the following form [A 310]:

$$\Lambda = \Lambda_0 - \left[\frac{eK}{300} \cdot \frac{(z^+ + z^-)F}{6\pi\eta^0} \right] + \left(\frac{e^2\chi}{6DKT} \Lambda_0 w \right) \qquad (1\text{-}61)$$

where χ is the effective radius of the ionic atmosphere, w is a function of ionic charge mobility, η is the viscosity, and D is the dielectric constant. The first term in brackets represents the electrophoretic effect and is dependent on the viscosity; the second term is due to the relaxation effect and is dependent on the dielectric constant. The question of the valid use of the bulk dielectric constant for a molecular-scale process has already been discussed.

According to theory, the effective radius K of the ionic atmosphere is related to the square root of the concentration of the ions and the dielectric constant of the medium. Accordingly, equation (1-61) becomes [A 310]:

$$\Lambda = \Lambda_0 - \left[\frac{29.15(z_+ + z_-)}{(DT)^{\frac{1}{2}}\eta} + \frac{9.90 \times 10^5}{(DT)^{\frac{3}{2}}}\Lambda_0 w\right][C(z_+ + z_-)]^{\frac{1}{2}} \quad (1\text{-}62)$$

A number of attempts have been made to extend this model to describe the effect of viscosity on conductivity at high solute concentrations. However, most treatments have fallen short in describing the phenomena. The following equation has had some success at higher concentrations (many moles/1iter), i.e., the data give a fit within a few percent [A 150]:

$$\Lambda(\eta/\eta^0) = \left[\Lambda_0 - \frac{B_1(C)^{\frac{1}{2}}}{1 + \chi a}\right]\left(1 + \frac{\Delta X}{X}\right) \quad (1\text{-}63)$$

Except for the relative viscosity term, this is another form of the Onsager equation. Although the equation has little theoretical justification at high concentrations, it nevertheless may be regarded as a phenomenological description of the data [A 150]. At low values of concentration, $\eta = \eta^0$ and the equation for conductivity reduces to the Onsager equation given previously. An idea of the magnitude of this correction term (η/η^0) is given in Table 1-V. Thus, the effect of η_{rel} on the conductance of concentrated propylene carbonate electrolytes is pronounced. In the case of the $ZnCl_2$ solution listed, the measured conductivity is lowered by $\sim 10^3$.

It has been observed that the addition of a low-viscosity solvent to a viscous electrolyte does indeed improve the measured conductivity.

TABLE 1-V

Relative Viscosities

Electrolyte	η/η^0	Reference number
1 M LiCl in H_2O	1.12	
4 M LiCl in H_2O	1.61	
9 M LiCl in H_2O	4.4	
0.02 moles KI/100 g PC	1.6	[A 151]
0.0875 moles KI/100 g PC	155	[A 151]
0.243 moles LiBr/100 g PC	45	[A 151]
0.528 moles $ZnCl_2$/100 g PC	900	[A 151]

An example of this is shown in Table 1-IV for the addition of ethylether (EE) to a solution of $LiAlCl_4$ in propylene carbonate (PC). Similar results were obtained with benzene and toluene additions [A 252].

This explanation (*i.e.*, viscosity change) cannot be the entire story, since a high conductivity is also obtained for $LiAlCl_4$ in the ethylether alone, a low-dielectric-constant solvent. Consideration of the chemical interactions between ion and solvent are also apparently necessary.

Stearn–Eyring Treatment. A number of recent discussions of ionic conductance (*e.g.*, [A 124,351]) have revived the model originally suggested by Stearn and Eyring [A 350]. This treatment is a formal development of the notion that the migration of an ion involves a succession of "jumps" from one equilibrium position to another in the liquid across a saddle-point potential energy configuration which is comparable to the activated step in a chemical reaction. The basic equation is as follows:

$$\lambda_i^0 = (ze_0 F/6h)L^2 \exp(-\Delta\mu_0^{\ddagger}/RT) \tag{1-64}$$

where ze_0 is the charge of the ion; L is the distance between equilibrium positions; and $\Delta\mu_0^{\ddagger}$ is the change in chemical potential required to reach the saddle point. By appropriate differentiation, one derives activation parameters for the migration process, *viz.*,

$$RT^2(\partial \ln \lambda_i^0/\partial T)_P = 2RT^2(\partial \ln L/\partial T)_P + (\Delta\bar{H}_0^{\ddagger})_P = E_p \tag{1-65}$$

$$RT^2(\partial \ln \lambda_i^0/\partial T)_V = 2RT^2(\partial \ln L/\partial T)_V + (\Delta U_0^{\ddagger})_V = E_v \tag{1-66}$$

and

$$-RT(\partial \ln \lambda_i^0/\partial P)_T + 2RT(\partial \ln L/\partial P)_T = \Delta V^{\ddagger} \tag{1-67}$$

where $(\Delta\bar{H}_0^{\ddagger})_P \cong E_p$ is the enthalpy of activation at constant pressure; $(\Delta\bar{U}_0^{\ddagger})_V$, which equals E_v since $(\partial \ln L/\partial T)_V$ is presumably zero, is the energy of activation at constant volume; and $\Delta V^{\ddagger}$ is the volume (change) of activation.

It was shown [A 351] that E_p and E_v must be carefully distinguished, not only because the latter is more satisfactory theoretically, but also because it is considerably different in magnitude from E_p. In addition, E_v is a simpler parameter, since it depends on the density of the solvent only and is independent of the temperature. It was also demonstrated [A 351] that an analysis of E_v may lead to an understanding of ion-solvent interactions since E_v is a sensitive function of the solvent density (packing).

The repeated finding that semilogarithmic plots of Λ_0 and $1/T$ accurately represent the temperature coefficient of ionic conductance shows that the use of transition-state theory as a starting point in the description on ionic mobility is at least justifiable from an experimental viewpoint.

A considerable literature has grown up on the interpretation of the energy parameters E_p and E_v for liquid viscosities, and similar considerations would apply for the conductance. Eyring suggested that E_v represents the energy for the migrating species to jump into a prepared vacancy, and that E_p [or rather $(\Delta \bar{H}_0{}^{\ddagger})_P$] is the sum of E_v and the energy required to prepare the vacancy. This suggestion, which has received much attention, implies that a vacancy or hole has essentially molecular dimensions.

However, it has been shown recently that this notion of a simple mechanistical distinction between E_p and E_v is not readily justified by experiment. Data for DMF suggest a relationship of the following form [A 124]:

$$E_v = B'/(V - V_0)$$

where B' and V_0 are constants and V_0 can be related to the volume *not available* for ion movement, *i.e.*, the closest-packed volume of the solvent. This is similar in concept to the van der Waals equation of state for gases. Indeed, it is observed that V_0 is independent of solute and does approach the computed value for closed packed DMF.

It is not possible with the limited range of data in the literature to express the variation of E_v with ion size quantitatively. However, we may note that: (1) at a given density, E_v is higher for smaller ions; and (2) E_v increases more rapidly with increase in ion size when the ion size is smaller, and is somewhat more sensitive to ion size at larger densities.

A problem in the theory is the development of the pre-exponent term in the Arrhenius-type representation:

$$\log \Lambda_0 = \log A - E_v/2.303\ RT \qquad (1\text{-}68)$$

The limited data on DMF indicate an equation of the form

$$A = rV^{2/3}(V - V_0'')$$

where
$$V_0'' \cong 44.3 + 3.6 \times 10^{-2} V_{salt}$$

The range of V over which this expression for A is known to hold is small, and the equations can be regarded only as tentative; however, the form is similar to that proposed earlier for a model involving a redistribution of free space as a necessary step before a jump may occur. In addition, the value of V_0'' (similar to the closepacked volume) is reasonable, and its dependence on ion size is in the correct sense.

Also calculable from this treatment is the volume of activation $\Delta V^\ddagger$. To a first approximation, $\Delta V^\ddagger$ may be assumed to comprise two parts—the volume of activation required to create a new vacancy or equilibrium position $\Delta V_v^\ddagger$ and the volume of activation required to allow a jump from one equilibrium position to another $\Delta V_j^\ddagger$.

"It appears that, although the form of the temperature coefficient of conductance is given by the transition-state theory, the quantitative prediction of experimental results from the theory, insofar as it may be attempted, is poor. ... The development of similar equations suggests the possibility of separating the contributions of ion and solvent parameters to the ionic environment and of developing a more satisfactory description of the limiting ionic mobility than the Walden rule" [A 124].

FARADAIC INEFFICIENCIES

The mathematical treatments developed above describe simple, well-defined electrode and mass-transport processes. The behavior of real systems can depart significantly from the ideal cases described. A number of the more important processes inhibiting the actual use of the available active material are described below.

It is convenient, for purposes of discussion, to define two broad classes of current inefficiencies—mass balance and kinetic effects. The current efficiency is incorporated into the figure-of-merit expression[1] via
$$K_1 = \epsilon_I \cdot \epsilon_v$$

where ϵ_v represents the voltage efficiency and ϵ_I the current, or Faradaic, efficiency.

[1] For further detail on the figure of merit, see the preface to this volume, p. viii.

Mass Balance

The mass-balance current inefficiencies involve the incomplete use of active material through the following processes: (1) parasitic processes, e.g., corrosion and solubility; (2) incomplete discharge reaction; and (3) insufficient loading of one or more reactants.

For example, the fraction of material lost via direct corrosion of the active material by electrolyte $\epsilon_I{}^c$ can be represented by the following equation:

$$\epsilon_I{}^c = \frac{w_{act} - (r_c \times t)}{w_{act}} \tag{1-69}$$

where r_c is the corrosion rate, w_{act} is the weight of active material, and t is the time.

Active material can also be lost by chemical dissolution and subsequent diffusion out of the catholyte or anolyte. The exact magnitude of this effect in controlling ϵ_I is determined, in part, by the diffusion constant of the material, the solubility of the active material in the battery electrolyte, and the diffusion tortuosity of the separator membrane. It is often stated that some solubility is desirable, since it provides a transport mechanism for nonconductive depolarizers to a conductive grid. In fact, the effectiveness of a number of organic depolarizers has been correlated on this basis. However, a high solubility of reactant can adversely affect a long life on wet stand.

It is necessary to compare the actual discharge stoichiometry with the theoretical stoichiometry used on computing Q_0. In other words, what is the actual number of electrons transferred per molecule of material undergoing discharge, compared to that required for complete oxidation or reduction? This problem is particularly pertinent to positive materials. For example, it is found that m-DNB in acid discharges at approximately 12 electrons per molecule, while in neutral solution 6–8 electrons per molecule are realized. The same effect is noted for the reduction of nickel oxide. As will be shown, the reduction of V_2O_5 in fused salt media involves substantially less charge than that required for reduction to the metal. To account for this stoichiometry effect, we introduce the following parameter:

$$\gamma_{st} = \frac{n_e{}'}{n_e}$$

where $n_e{}'$ is the actual number of electrons released per molecule discharged and n_e is the number of electrons on which the theoretical cell reaction (Q_0) is based. Note that this quantity is subject to change by alteration of the experimental conditions; *e.g.*, changes in pH are effective in changing γ_{st}.

Another mass-balance inefficiency involves the coulombic mismatch of active plates. For many cell reactions, it is also necessary to consider the electrolyte as one of the reactants (as opposed to solvating reaction product). For example, the complete discharge of m-DNB in neutral aqueous systems proceeds by the following equation:

$$\phi(NO_2)_2 + 12e^- + 8H_2O \rightarrow \phi(NH_2)_2 + 12OH^-$$

Electrolyte can become the material present in least amount and, thus, can determine the excess amount of active plate material. Of course, this loss of electrolyte by a cell reaction will have other effects, such as changes of electrolyte conductivity and salting out of product.

Kinetic Effects

The plate discharge processes are terminated when the chosen cutoff voltage E_c is reached; the material not discharged when E_c is reached represents a loss in current efficiency. In effect, choosing a cutoff voltage defines the voltage efficiency; ϵ_I thus becomes the dependent variable.

The question arises, what specific processes will bring about changes with time in the activation polarization η_{act}? For example, involved in η_{act} is a possible change in discharge mechanism as the plate materials are consumed. An example of this is observed in the slow discharge of AgO to silver. Rather than proceeding directly to the metal, a stable intermediate oxide, Ag_2O, is first formed. At the potentials where the discharge of AgO is fast, the subsequent discharge of Ag_2O is slow. Complete discharge to silver metal is obtained only by a further increase in η_{act}. Only by initially polarizing the cell to a voltage where both reactions are fast can a single discharge plateau be observed, and η_{act} is constant with time. Constancy of discharge can be achieved by eliminating the higher oxidation state. However, this decreases E_{cell} and the theoretical energy density of the material.

The initial electrical conductivity of negative plates is generally quite high, so that its contribution to the cell ohmic losses can be quite low. With positive plates, the ohmic term is of more significance because of the higher resistance of the compound depolarizers; at high drain rates, this factor can dominate in determining energy density.

The initial η_Ω is generally considered in the voltage efficiency term, since it influences the average operating potential. The influence on ϵ_I comes about as η_Ω changes with time.

As mentioned previously, cell reactions can involve the direct consumption of electrolyte, and, as the reaction proceeds, electrolyte conductivity will drop. Alternatively, the electrolyte may not be consumed directly, but may be used to solvate reaction product. This will also change conductivity. In either case, η_Ω is a function of the number of ampere-hours generated.

A second ohmic effect involves the resistive losses within the battery plates and the changes of plate resistance as the discharge proceeds. As will be shown, the change in plate resistivity can be either positive or negative. For example, a decrease in resistivity is achieved when plate discharge produces a more conductive material, *e.g.*,

$$AgCl + e^- \rightarrow Ag^0 + Cl^-$$

If the depolarizer is an electrical insulator, a conductive binder must be included within the battery plate structure to provide a means of withdrawing electrical power. The electric charge is transferred from a conductive particle (*e.g.*, carbon) into the nonconducting depolarizer. The depth to which this charge can penetrate into the particle can determine the efficiency of use of the active material, particularly at high current densities.

Consider a depolarizer coating of 1-cm² area and thickness t on a conductive substrate. The current density (in mA/cm²) is described by the following expression:

$$CD = \sigma E/t$$

where E is the potential drop across the film and σ is the conductivity of the film. Under the restraint that $E = 100$ mV, $\sigma = 10^{-8}(\Omega\text{-cm})^{-1}$, and $CD = 10$ mA/cm², the thickness (in cm) is

$$t = \frac{(10^{-8})(10^2)}{10} = 10^{-7}$$

Consider next a more conductive film ($\sigma = 7 \times 10^{-2}$); the thickness (in cm) is

$$t = \frac{(7 \times 10^{-2})(10^2)}{10} = 0.7$$

It would be expected that a greater use efficiency would result from the more conductive film, particularly at higher discharge rates.

The most difficult features to describe analytically are the changes in mass-transport processes. As with η_Ω, the initial factors are considered principally in the ϵ_V term. The influence of ϵ_I comes about only as η_c changes. As a plate discharges, reaction product is injected into the electrolyte, and often material is removed from the electrolyte. The product will either form a precipitate or will dissolve within the electrolyte. The amount of material (solvated or otherwise) in the vicinity of the electrode will affect the subsequent mass transport of reactants to the electrodes. In some batteries (*e.g.*, $Pb/H_2SO_4/PbO_2$), a precipitate is formed immediately, while in others (Mg/H_2O) a precipitate is formed only after a substantial portion of material has discharged. It is readily apparent that a precipitate on the surface of an electrode will screen a significant portion of active material from electrolyte. The same effect will result from gels formed within the electrode pores. The question of the amount of the active material blocked is tied up with the pore structure and surface area of the plate and the porosity of the precipitate.

Note that the pore structure of a plate need not be constant throughout a discharge, particularly when there is a difference in density between active material and product.

WEIGHT EFFICIENCY K_2

The remaining factor which is necessary to describe the practical energy density of a battery is the weight factor K_2. This term is defined as the ratio of the weight of the redox couples w_0 to the overall weight ΣW of the battery.

In analyzing ΣW, there are three dominant terms to consider— w_0, w_e, and w_s. The weight of the redox couples w_0 is easily calculated. The weight of the electrolyte w_e depends upon the choice of electrolyte composition, the specific cell configuration, and the porosity of the electrodes; the weight of the basic structural components of the battery (current connectors, case, *etc.*) w_s depends upon the composition of these components and the cell configuration. For each cell configuration, it should also be possible to define an irreducible set of volumes, V_e and V_s. When multiplied, respectively, by the density of the electrolyte ρ_e and the average density of the structural components ρ_s, V_e and V_s will yield the minimum values of w_e and w_s, respectively.

Many of the special requirements for high-energy batteries can be translated into extra terms in ΣW. Thus, for example, a requirement of high reliability can be translated into additional weight, since overdesign is the easiest way of ensuring reliability. The extent of overdesign will depend on the complexity of the system and on the known, or likely, modes of failure. The additional weight will include the activator system which varies for different systems.

As a further example, a requirement of an extended temperature range can also be translated into terms of additional weight. Systems whose range of applicability is within the stipulated temperature limits will require no additional weight. On the other hand, if the low-temperature extreme is below the freezing point of the electrolyte compatible with a particular system, provision for auxiliary heating (electrical or chemical) must be made, at least to activate the battery. High temperatures may require additional components, e.g., heat exchangers. The resultant changes in weight can be computed in principle for each case.

In summary, the total weight of the battery and its associated components can be written as

$$\Sigma W = w_0 + w_e + w_s + w_r + w_t + w_m$$

where w_r and w_t are the additional weight increments due to special requirements of reliability and temperature control and w_m is the sum of similar increments made necessary by other special features, such as activating systems.

By similar reasoning, it is possible to evaluate a volume figure of merit, M_v, which gives the practical expectation for W-hr/cm^3 by dividing each of the weight terms in the expression for M_w by the corresponding densities. It follows that

$$M_v = K_1(Q_0\rho_0)(w_0/\rho_0)/V$$

where

$$V = \frac{w_0}{\rho_0} + \frac{w_e}{\rho_e} + \frac{w_s}{\rho_s} + \frac{w_r}{\rho_r} + \frac{w_t}{\rho_t} + \frac{w_m}{\rho_m}$$

where the ρ_i's are the relevant density values.

PERFORMANCE COMPARISON

A few examples of the experimental performances of a number of battery systems are listed in Table 1-VI according to their electrochemical and weight efficiencies. This listing is by no means all inclusive; as will be shown, the observed energy density of a battery is a function

TABLE 1-VI

Examples of Battery Efficiencies

Battery	Reference number	Q_0	ϵ_I	ϵ_V	K_2	M_w (W-hr/lb)
Pb/PbO$_2$	[A 1]	112	0.35	0.96	0.41	14.7
Cd/NiO	[A 3]	107	0.39	0.925	0.39	15
Zn/AgO	[A 9]	254	0.75	0.766	0.45	66.1
Li/CuF$_2$*	[A 85]	754	0.425	0.9	0.268	76
N$_2$H$_4$/O$_2$	[A 302]	1190	1.0	0.58	0.125	87
Zn/AgO†	[A 143]	254	0.75	0.766	0.071	9.8
Mg/NH$_3$/m-DNB†	[A 144]	855			0.035	10

* Experimental battery.
† Reserve battery.

of size, rate, and duty cycle. The data given for the first three systems are for deep discharge of a secondary battery at the 20-hr rate. Two high-rate reserve batteries and a fuel cell system are included for comparison.

As shown, the performance of reserve batteries is seriously inhibited by a high system weight. It is apparent that seldom does the weight of the active plate material account for half the battery weight, and that seldom is the effective current efficiency greater than 75%. The data given for the Li/CuF$_2$ battery is for a laboratory cell discharge at a very low rate, while the data for the N$_2$H$_4$/O$_2$ cell describes a 6-kW system operating for 42 hr at a current density of 150 mA/cm^2.

The question to be asked is whether each of these terms represents the optimum possible for a given application. The chemical and physical factors which determine the magnitudes of Q_0, K_1, and K_2 are discussed in the remainder of this book.

Chapter 2

Electrochemically Active Materials Q_0: Aqueous Electrolytes

In the preceding chapter, the extent to which a general theory exists for the description of the electrode and solution processes involved in the discharge of battery couples was discussed. The items next to be considered in describing the present status of high-energy batteries are the specific electrode and electrolyte materials. This subject is rather extensive and is subdivided according to the type of electrolyte employed, *i.e.*, aqueous (Chapter 2), nonaqueous, inorganic (Chapter 3), and nonaqueous, organic (Chapter 4).

The thermodynamic energy capacity Q_0 of a number of possible electrode materials is presented. A brief review is also given of the characteristics of these materials in working cells.

ELECTROLYTES

For minimal ohmic resistance in an aqueous electrolyte, an acid is desirable. However, only a few plate materials are stable in such an environment. The discharge voltage and rate of discharge of lead and lead dioxide are acceptable; the high equivalent weights are the principal drawbacks.

Aqueous KOH is a highly conductive electrolyte and is compatible with a greater variety of materials. A compilation of some physical properties of this solvent is given in Tables 2-IA, 2-IB, and 2-IC. The use of quaternary ammonium electrolytes instead of potassium hydroxide or the hydroxides of other alkali metals has been suggested [P 425]. These substances are miscible with water and form stable electrolyte solutions which are very strong bases; the conductivity of these solutions is similar to that of conventional alkali hydroxide

TABLE 2-IA

Electrical Resistance of Potassium Hydroxide Solutions*

Concentration of KOH (wt.%)	Resistance (Ω/cm³)			
	18°C	25°C	75°C	150°C
2.37		10.63	5.89	
5	5.40			
5.73		4.47		
6.37		4.10	2.27	
9.72		2.83	1.58	
10	3.20			
10.67		2.63	1.42	
12.78		2.28		
15	2.34			
19.38		1.75		
20	2.00			
20.34		1.61		0.53
22.77			0.83	
25	1.86			
25.90		1.57		
27.84		1.57	0.78	
29.38		1.54	0.75	
30	1.84			
33.56		1.60	0.84	
35	1.96			
39.08		1.84	0.98	
40	2.20			
41.25				0.41
42.09		1.91		0.40
45	2.56			
49.16		2.55		
54.17				0.46
59.36				0.53
60.10			1.34	

* Taken from [A 1,372,374].

electrolytes. It has been suggested that the use of quaternary ammonium electrolytes could result in batteries having better wet-stand life, since the effect of these electrolyte solutions is to greatly increase the hydrogen overvoltage of the electrodes of alkaline storage battery systems. Further, the use of these electrolytes is reported to improve the recharge characteristics of the negative plates [P 425].

TABLE 2-IB
Solubility of Potassium Hydroxide Solutions*

Temperature (°C)	Solid-phase composition	Solubility Molality (moles KOH/1000 g H_2O)	KOH (wt.%)
−45	KOH · $4H_2O$	10.2	36.4
−40	”	11.0	38.2
−35	”	12.2	40.6
−33	Transition point	14.44	44.7
−20	KOH · $2H_2O$	15.4	46.4
0	”	17.3	49.3
10	”	18.4	50.8
20	”	19.9	52.8
30	”	22.5	55.8
33	Transition point	24.0	57.4
40	KOH · H_2O	24.4	57.8
60	”	26.0	59.3
80	”	28.4	61.4
100	”	31.6	63.9
120	”	36.2	67.0
140	”	45.6	71.9

* Taken from [A 373].

TABLE 2-IC
Viscosity of Potassium Hydroxide Solutions

Concentration of KOH Wt.%	Mol.%	Viscosity (cP) 18°C*	25°C†	75°C†
5		1.17		
10		1.30		
10.67	3.70		1.13	0.513
15		1.48		
20		1.72		
22.77	8.65		1.64	0.712
25		2.05		
27.84	11.03		1.98	0.816
29.38	11.80		2.08	0.913
30		2.50		
39.08	17.09		3.29	1.27
49.16	23.70		6.63	2.22
54.30	27.62			2.98
60.10	32.62			4.78

* Taken from [A 1].
† Taken from [A 372].

The less corrosive solution of $ZnCl_2$ plus NH_4Cl is used as the electrolyte for the low-energy-density LeClanché cell. Details concerning this system may be found in the work of Vinal [A 200].

INORGANIC NEGATIVES

The negative plate materials most commonly used with an alkaline electrolyte are iron, cadmium, and zinc. Even though iron is relatively inexpensive, there are a number of electrochemical difficulties which complicate its use. Cadmium is a more effective material for secondary cells, having a high discharge rate, an insoluble reaction product, $Cd(OH)_2$, and a negligible corrosion rate.

The most widely used negative material is zinc, which is employed in the Reuben cell, the LeClanché cell, the alkaline LeClanché cell, and the "silver–zinc" cells. This material is more active than cadmium, has a lower equivalent weight, and is cheaper. A corrosion problem exists, although this can be minimized by amalgamating the electrode surface. A summary of equilibrium potentials–pH data for many common plate materials is given in the work of Pourbaix *et al.* [A 299].

Zinc

The equivalent weight of zinc is 32.61; the standard potential of the Zn/Zn^{+2} couple is [A 8]:

$$Zn^{+2} + 2e^- \leftrightarrows Zn \qquad E^0 = -0.763 \text{ V}$$

In alkali, the reaction is:

$$Zn(OH)_2 + 2H^+ + 2e^- \rightleftarrows Zn + 2H_2O \qquad E^0 = -1.245 \text{ V}$$

This electrode is unstable toward electrolyte decomposition; however, the rate of hydrogen evolution is low; a further increase in overvoltage can be accomplished by amalgamating with mercury. (Details concerning such parasitic reactions and their prevention are discussed elsewhere in this book.)

The electrochemistry of zinc in alkaline solution is complicated by the formation of ZnO_2^{-2}. The solubility product of zinc hydroxide [A8]

$$Zn(OH)_2 \rightarrow ZnO_2^{-2} + 2H^+$$

is $K_{sp} = [ZnO_2^{-2}][H^+]^2 = 1 \times 10^{-29}$. In neutral solutions, the saturation concentration of zinc is 10^{-15} M. However, in concentrated alkali, e.g., 6 N KOH, where the hydrogen ion concentration is 0.16×10^{-14} M, the saturation concentration of ZnO_2^{-2} is 3.6 M. Thus, with incorporation of this equation, in concentrated alkali, zinc oxidizes according to

$$ZnO_2^{-2} + 2H_2O + 2e^- \rightleftarrows Zn + 4OH^- \qquad E^0 = -1.216 \text{ V}$$

or [A 353]

$$Zn(OH)_4^{-2} + 2e^- \rightleftarrows Zn + 4 \text{ OH}^-$$

The discharge reaction is followed by a reaction of the following type:

$$Zn(OH)_4^{-2} \rightleftarrows ZnO + H_2O + 2 \text{ OH}^-$$

This equilibrium is only very slowly established, requiring, in some cases, many months [A 352]. This tendency for zincate solutions to become supersaturated is particularly detrimental to the operation of secondary batteries based on zinc negatives.

The behavior of a zinc electrode on discharge in caustic can be quite complex as illustrated by the following discussion of a potentiostatic study of zinc in KOH solutions (0.01, 0.1, 1, and 6N) [A 344]. "In all solutions of concentration above 0.1 N KOH, the electrode first shows an initial active region in which the current increases rapidly with voltage as it is made more positive from about -1.3 V. In this region, zinc actively goes into solution as zincate ions. There are only small differences between KOH and KOH saturated with ZnO in 6 N solutions; consequently, saturated zincate solutions can still actively dissolve zinc; ZnO and $Zn(OH)_2$ are presumably precipitated in the bulk of the solution later on. As far as the electrode surface is concerned, not much precipitation takes place in the ascending portion of the E–I curves. It seems, therefore, that the zincate ions are relatively stable under supersaturation conditions."

"The rate of anodic dissolution, however, does not continue to increase with increasing voltage, but soon reaches a maximum value. Although this maximum rate increases markedly with increasing KOH concentration, yet it is only a fraction of the limiting rate of OH⁻ ion diffusion through the solution, even in 6 N solutions. In this maximum rate region, the electrode becomes coated with a visible grey deposit. At a certain voltage, the current suddenly decreases, the length of the current plateau being dependent on the concentration of the solution."

"The OH⁻ tends to increase the rates of dissolution; however, the zincate ions exert a marked but variable influence of their own and thereby complicate the observed behavior. It seems there are at least two reactions at different anode voltages. One of these can be accelerated and the other retarded by zincate ions under certain conditions. This behavior of zincate ions appears to be connected with the varying degree of its preferential adsorption in different potential regions and concentrations."

"It also appears that the insoluble oxide formed at -1.1 to -1.0 V is gradually converted to a more soluble form and the current, after initially falling, increases to a steady-state value. At less negative potentials, the rate of oxide dissolution remains relatively constant over a considerable voltage range. It would appear that the behavior of the electrode in this potential region is governed by a fixed rate of dissolution of the oxide in the electrolyte, the potential increment merely increasing the thickness of the layer so as to maintain the field strength required to continue the necessary cation transport, by high field conduction for the fixed rate of dissolution" [A 344].

Magnesium[2]

The standard potential of the magnesium electrode is -2.43 V. However, the steady-state working potential is generally of the order of -1.5 V [A 192]. It is postulated that the magnesium electrode is in a "passive state" [A 293]; the rate of corrosion and discharge are controlled by a $Mg(OH)_2$ film.

This film leads to a long delay time (*e.g.*, 100 sec) in restarting a cell. To some extent, this activation time can be decreased by alloying magnesium with 0.1–0.7% zinc and 0.05–0.5% calcium [P 153]. The addition of 0.03–1.7% indium has also been indicated as useful [P 154]. Apparently, an alloy exists with a range of aluminium content of 1.6–2.5% and a coordinated zinc range of 0.8–2.5% in which there is a maximum in anode efficiency and a minimum in delayed-action time [P 144]. Supposedly, the hydrogen produced during corrosion of the magnesium develops a pressure which forces the electrolyte out of the MgO layer and out of the spacer immediately in contact with the anode, increasing the ohmic resistance. An anode design has been suggested which permits venting of the negative [P 155].

[2] See also Chapter 5, pp. 160–163.

It may be possible to improve on the capacity of the magnesium negative by alloying with 30–48% lithium. It has been shown that, in this range, the attack by water is $\frac{1}{10}$ that of lithium and the voltage is somewhat higher than with magnesium alone, as is the capacity. The stability in air is similar to that of magnesium [P 172].

A number of different aqueous inorganic electrolytes have been used in conjunction with magnesium anodes. An early patent [P 147] discussed an electrolyte of 30% chromate ion in 20–25% H_3PO_4 which reportedly achieved an efficiency of 84% at 3 mA/cm². The hydroscopic properties of phosphoric acid helped minimize the loss of water by evaporation. An electrolyte of better electrical conductivity included an alkali bromide plus an alkaline earth bromide [P 146]. A soluble chromate (0.05–2 g/liter) is added to inhibit corrosion; some ammonia is also added to make the electrolyte alkaline (pH 8.5). Ammonium bromide and chromate can be used alone [P 150]; however, aqueous $Mg(ClO_4)_2$ has a greater conductivity (e.g., [P 151]). The discharge of magnesium and zinc negatives results in an increase in resistance at the electrodes. The use of alkali sulfamates has been suggested to avoid this problem [P 141,152].

As indicated, magnesium has a potential sufficient to decompose aqueous electrolytes, i.e.,

$$Mg + 2H_2O \rightarrow Mg(OH)_2 + H_2 + heat$$

Besides being wasteful of active material, the heat generated will evaporate water from vented cells. The construction of sealed cells becomes difficult because of the pressure buildup. For low-temperature operation, however, this heating can be useful [P 145].

Aluminum

The half-cell reactions for the anodic discharge of aluminum are as follows [A 8]:

$$Al^{+3} + 3e^- \rightarrow Al \qquad E^0 = -1.66 \text{ V}$$
$$Al(OH)_3 + 3e^- \rightarrow Al + 3OH^- \qquad E^0 = -2.31 \text{ V}$$
$$(H_2AlO_3)^- + H_2O + 3e^- \rightarrow Al + 4OH^- \qquad E^0 = -2.35 \text{ V}$$

The equivalent weight of aluminum is 9, making it superior to magnesium from this point of view.

The problem is that aluminum forms an oxide film which, although it prevents extensive corrosion, also deactivates the discharge properties. One approach has been to employ an electrolyte of aqueous caustic containing a metal zincate. The aluminum oxide is soluble in this environment and the corrosion is not excessive, since, apparently, a coating of metallic zinc is formed by reaction with the aluminum [P 158].

By employing such an electrolyte and by surface-amalgamating the aluminum, it has been possible to obtain current efficiencies $\geqslant 60\%$ at current densities of 3.5–140 mA/cm². Organic materials, e.g., "Hyamine 3500," were also reported to be satisfactory inhibitors with unamalgamated aluminum. It was concluded [A 18] that optimum battery performance would result: (1) from the use of such inhibitors; (2) from the elimination of sulfides and other corrosion accelerators; (3) from the use of highly pure aluminum or specially resistant aluminum–magnesium or aluminum–zinc alloys; (4) with effective heat removal; (5) by operation at high current densities (preferably $\geqslant 10$ mA/cm²); and (6) with anode–electrolyte separation during periods of inaction.

Aluminum anodes have been suggested for reserve batteries employing a NaOH electrolyte [A 193]. A current efficiency of 99% was obtained at 100 mA/cm²; the efficiency dropped to 58% at 10 mA/cm².

Another approach has been to employ an aqueous electrolyte of AlCl₃ (pH 3.5–4). Chromate ion is included to restrict attack of the chloride on the anode. A long shelf life is reported [P 159]. Added stability to corrosion can be achieved by alloying the aluminum with 0.001–5% Zn, 0.02–0.5% Sn, 0.02–1% Bi, and 0.1–10% Mg [P 160].

An acidic (pH 5–5.6) electrolyte containing borax and manganous chloride has also been proposed [P 161] as a method of preventing film formation. Oxalic, acetic, sulfonic, nitric, or sulfuric acids can also be used as electrolytes, if chloride ion is included to break up the protective film [A 267].

It has been claimed that the discharge characteristics of aluminum are greatly improved by alloying with approximately 0.2% Sn [P 404]. Gassing is reduced and the electrode is capable of high current densities.

Titanium

The use of titanium as a battery negative has been complicated by similar protective film formation. A series of patents have described titanium alloys for which these problems reportedly do not exist. For example, an alloy of 50 wt. % Ti and at least 30 wt. % Mo is suitable as a negative in a concentrated caustic electrolyte. The addition of beryllium, aluminum, or boron increases voltage and capacity [P 162]. An alloy of titanium with 27% Mo and 10% Nb is reported extremely stable to gassing [P 163]. Similar results are found for alloys high in beryllium plus aluminum [P 164] and vanadium plus chromium [P 165]. Open circuit potentials, rates of corrosion, and load data were given, and it was stated that the alloying constituents were also consumed. However, cell data in units of W-hr/lb or A-hr/lb of electrode were not available.

Another approach involves the use of aqueous fluoride media for battery discharge. The fluoride salts of titanium are soluble; consequently, titanium exhibits satisfactory electrode characteristics in dilute HF containing added NH_4F[A 46]. It was determined [A 46] that the value for the average equivalents change for oxidation of titanium and zirconium anodes was 4.

Even in fluoride solutions, passivation occurs at high current densities. The critical passivation current i_p in 1 N HF is 110 mA/cm²; in 0.075 N HF plus 6% NH_4F, i_p is 40 mA/cm². Zirconium shows less of a tendency to passivate, even in the absence of HF [A 46]. The nature of the surface film formed on titanium and zirconium in solutions containing HF and NH_4F was not ascertained. In a suggested battery electrolyte of 0.075 N HF, 5–6% NH_4F, and 0.1 N KCl, the corrosion rate of titanium was measured as 22 mdd (milligrams/square decimeter/day), which, while not excessive, sets an upper limit of about 6 months on the life of a 0.25-mm sheet anode [A 46]. The role of HF is to increase i_p, of NH_4F to inhibit corrosion [A 47], and of KCl to provide electrical conductivity.

Zirconium corroded rapidly in all acid fluoride solutions investigated [A 46], and the addition of NH_4F was of no benefit. The corrosion rate in the mixed electrolyte described above was 6000 mdd; it was 800 mdd in 8% NH_4F alone [A 46]. Small concentrations of lead, zirconium, and chromate ions reduced the corrosion rate in 0.1 N HF, but not to a useful level.

The discharge properties of titanium anodes in a variety of fluoride–containing electrolytes have been discussed [P 422]; working cells employing titanium anodes have been constructed [A 46]. A Ti/MnO_2 cell had an open circuit voltage of 1.75 V in the electrolyte described above. When discharged at 10 mA/cm², the voltage remained at about

1.4 V for delivery of 25 A-min and dropped to 1.0 V at 39 A-min. This corresponded to 81% utilization of the anode. Fluoride solutions apparently affect MnO_2 electrodes deleteriously, so that even the open circuit voltage of such cells falls to zero after 2 or 3 days.

Ti/PbO_2 cells have a higher initial open circuit voltage (2.2–2.7 V); a cell potential of 1.95 V at 2 A/in.^{2} has been reported [P 422]. The decrease in voltage during discharge of these cells is due to dissolution of the PbO_2 electrode in fluoride solutions. The addition of a small amount of H_2SO_4 increased the cell voltage, probably because of an increase in the stability and potential of the PbO_2 electrode. However, the titanium anode polarized rapidly during continuous discharge in solutions containing H_2SO_4.

PbO_2 has good discharge characteristics in reserve batteries employing fluoroboric acid as an electrolyte. If it is assumed that titanium can be discharged satisfactorily in such solutions, a Ti/PbO_2 couple would provide a reserve battery with a Q_0 value of 230 W-hr/lb compared to 112 W-hr/lb for Pb/PbO_2.

Indium

The equivalent weight of indium is 38.2. A couple of In/AgO has a theoretical capacity of 200 W-hr/lb, which is higher than that for Cd/AgO (148 W-hr/lb), but not as high as for Zn/AgO (248 W-hr/lb). The principal advantage of the indium system is that it does not evolve gas on storage or on discharge. Indium anodes are, therefore, useful for primary batteries; they are unsuitable for secondary batteries because of difficulties in recharging.

Indium is a soft metal which has poor mechanical properties, such as elastic limit, ultimate tensile strength, and bending strength. These features complicate the construction of large, unsupported electrodes. Furthermore, trace quantities of impurities in commercial indium sometimes have an adverse effect on the electrochemical characteristics of the cell when first assembled. Most of these problems can be solved by alloying indium with about 2% bismuth [A 45]. Indium alone, in an alkaline electrolyte, will support a current of only 0.7 mA/cm^2, but the alloy will sustain 20 mA/cm^2. Furthermore, the alloy recovers more rapidly from temporary overloads. The corrosion rate of indium–bismuth alloys does not differ from that of indium, so that the advantage of long-term stability is unaffected [P 363,364].

A comparison with amalgamated zinc electrodes at low temperatures [A 26] is shown in Fig. 2-1. In the temperature range of −8 to 0°C, the In/Bi electrode is more effective, yielding at least 70% of its capacity at current densities of 2.5 mA/cm².

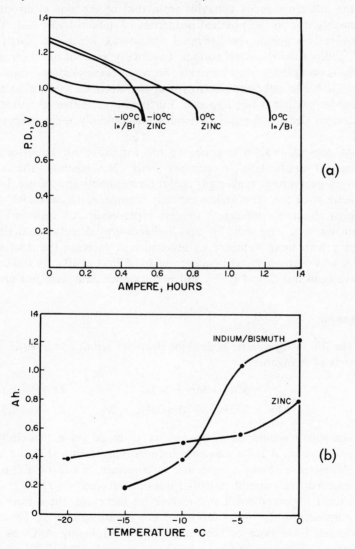

Fig. 2-1. (a) Discharge curves of indium–bismuth and zinc anodes at a current density of 20 mA/in². (b) Change in capacity with temperature. Current density at anode is 20 mA/in.² [A 26].

Anodic oxidation of indium or of indium–bismuth alloys in alkaline solutions yields a hydroxide having a formula closely corresponding to $In(OH)_3 \cdot 6H_2O$. The high efficiency of the In/Bi electrode and the difference in its behavior from that of the zinc electrode is presumably due to the physical properties of indium hydroxide. This hydroxide is a gelatinous, hydrated compound, which, at first, falls away readily from the metal surface. Eventually, most of the electrolyte volume is occupied by the hydroxide; however, because of its gelatinous nature, it holds sufficient electrolyte at the electrode surface so that the anode reaction is not impeded. Furthermore, neither temperature nor current density appears to modify substantially these properties [A 26].

At present, indium is probably too expensive for large-volume commercial application in primary cells. Nevertheless, the high hydrogen overvoltage makes gas evolution negligible and, hence, leads to a long shelf life. The higher capacity, compared to cadmium, and the high discharge efficiency suggest replacement of cadmium by indium where a long shelf life and higher energy density are desired. Alloying with light metals, e.g., lithium, may increase the discharge voltage and decrease the equivalent weight. Alloys of 10% Li and 0.1% Bi have been mentioned [P 363], but performance data were not given.

Manganese

The following equations describe the equilibrium anodic discharge reactions of manganese:

$$Mn \rightarrow Mn^{+2} + 2e^- \qquad E^0 = -1.18 \text{ V}$$
$$Mn + 2\,OH^- \rightarrow Mn(OH)_2 + 2e^- \qquad E^0 = -1.55 \text{ V}$$

The equivalent weight of manganese is 27, based on a two-electron change. As such, it is intermediate between magnesium (12) and zinc (32). Manganese shows a tendency to passivate in caustic solution; the observed open circuit potential does not exceed -1.11 V.

Except for some work in the Russian literature, there has been little emphasis on the use of this metal as a negative material. The cells constructed have been of the reserve type, employing AgCl as the depolarizer. Seawater has been used as the electrolyte [P 399]; it has also been reported that an aqueous solution of $CaCl_2$ (15–16%) and $MgCl_2$ (9–10%) at pH 2–4 increases the anode life and anode potential

[P 400]. A reserve battery containing a concentrated $HClO_4$ electrolyte has also been described [P 412]. The electrochemical properties of manganese and its compounds were discussed at a recent symposium [A 211].

Chemisorbed Hydrogen

Durable negative plates for alkaline storage batteries reportedly have been prepared from hydrogen-containing Raney nickel (RNi). The discharge reaction is apparently

$$RNiH + OH^- \rightarrow RNi + H_2O + e^-$$

A cell was formed with a positive of nickel oxide and an electrolyte of 20–30% KOH. The observed potential was 1.3–1.4 V. The capacity of the negative was 200–400 A-hr/kg RNi [P 413].

INORGANIC POSITIVES

Unless otherwise noted, the depolarizers discussed below are intended for primary cells. Most of these materials are nonconducting salts, so that an electrically conducting material must be provided; *e.g.*, the cathode mix could be formed by mechanically blending the depolarizer with a carbon, such as acetylene black. The blended mix could be employed in a bobbin, as is done with the conventional LeClanché cells. Another effective structure is a plastic-bonded plate, whereby a binder, such as polyvinyl acetate, is included in the mix and the product is attached to a metal sheet [P 113].

For purposes of comparison, a number of battery couples are considered with zinc anodes; the theoretical capacity Q_0 (in W-hr/lb) is given in Table 2-II. The use of a zinc anode rather than magnesium avoids the problem of deciding on a realistic voltage and capacity for the magnesium electrode in an aqueous electrolyte. The data in Table 2-II, however, also reflect the normalizing contribution of zinc. This problem can be shown by the following specific exercise. For example, it is desirable to increase capacity by replacing silver positives with a material of lower equivalent weight, such as $LiMnO_4$ or m-DNB. How much of an improvement could result, ignoring the effects of cathode voltage? Substitution of $LiMnO_4$ for AgO improves capacity by a factor of 1.98 when coupled with magnesium, by a

TABLE 2-II

Comparison of Inorganic Depolarizers

Couple	Energy density (W-hr/lb)	Energy-density ratio (normalized to Zn/m-DNB)
Zn/m-DNB	330 ($E_C^0 = +0.02$)	1
Zn/LiMnO$_4$	230	0.7
Zn/Li$_2$CrO$_4$	165	0.5
Zn/LiBiO$_3$	135	0.4
Zn/Li$_2$S$_2$O$_8$	116	0.35
Zn/Li$_2$O$_2$	280	0.85
Zn/LiNO$_3$	215	0.65
Zn/LiNO$_2$	108	0.33
Zn/LiBrO$_3$	356 ($E_{BrO_3^-}^0 = +0.6$)	1.16
Zn/Hg(BrO$_3$)$_2$	293	0.89
Zn/LiIO$_3$	260	0.79
Zn/Ca(IO$_3$)$_2$	272	0.82
Zn/LiIO$_4$	310 ($E_c = +0.26$)	0.91
Zn/LiIO$_4$	400 ($E_c = +0.70$)	1.22
Zn/AgO	238	
Zn/S	185	0.56
Zn/Br$_2$	176	
Zn/AgCl	100	
Zn/CuCl	127	

factor of 1.45 when coupled with cadmium, and by a factor of 0.8 when coupled with zinc. Thus, in this sense, there exists some lack of rigor in comparing materials for either positives or negatives.

Furthermore, it must be recognized that highly active positives, like active negatives, will "corrode" and a theoretical potential in excess of the decomposition potential of the electrolyte will not necessarily be observed.

H$_2$O/H$_2$

Highly active anode materials, e.g., magnesium, have been formed into batteries in which the cathode reaction is the reduction of water:

$$H_2O + e^- \rightarrow \tfrac{1}{2} H_2 + OH^-$$

Such systems employ inert carbon cathodes, although low-hydrogen-overvoltage materials, such as nickel, cobalt, and iron, are more effective [P 111]. The accumulation of hydrogen bubbles can limit access of the electrolyte, polarizing the cell; this problem is obviated by proper design of the electrode structure [P 112]. An open circuit voltage of >1 V was observed for a platinum-black cathode combined with a magnesium anode. At this voltage, the theoretical capacity is 410 W-hr/lb. There are, of course, a number of problems involved in achieving a high performance in practice. One of the more important is disposal of the hydrogen produced. For every 18 g of water consumed, 11.2 liters of hydrogen (STP) are generated.

O_2/OH^-

The low-current-drain, air-depolarized battery is a common device, generally consisting of a zinc anode and an exposed catalyzed carbon cathode [P 134,140,148]. At the equilibrium potential, a Zn/O_2 cell has a theoretical capacity of 480 W-hr/lb. However, the oxygen reaction, even on a platinum catalyst, is slow; on load and with an alkaline electrolyte, the cathode potential falls from approximately +0.4 to approximately +0.1 V (*versus* the normal hydrogen electrode).

This cell can be compared with the chlorine-depolarized cell described below. If the cell polarization at low loads is taken into account, both systems have essentially the same expected capacity, so that a choice between them cannot be made on the basis of thermodynamic calculations alone. The problems involved in the construction of a gas-depolarized cathode are discussed in the section dealing with the Cl_2/Cl^- couple.

H_2O_2/OH^-

Hydrogen peroxide has been employed as a battery positive [P 118]. Peroxide cathodes also have been used in conjunction with some fuel cell systems [A 21]. Unfortunately, the high theoretical potential of peroxide in alkaline electrolyte is not achieved in practice. The observed potential is a mixed potential, the actual value depending on the particular catalyst–electrode used. The $Zn/KOH/H_2O_2$ system has a Q_0 value of 280 W-hr/lb if a cathode potential of +0.1 V is assumed.

In acid system, the equations have the following from:

$$H_2O_2 + 2H^+ + 2e^- \rightarrow 2H_2O \qquad E^0 = +1.77 \ \text{V}$$
$$H_2O_2 \rightarrow O_2 + 2H^+ + 2e^- \qquad E^0 = +0.682 \ \text{V}$$

A peroxide half cell operated in 3N H_2SO_4 has an open circuit potential of approximately $+0.5$ V and shows very little subsequent polarization. Since concentrated H_2O_2 is difficult to handle, the solid alkali peroxides can be considered.

Halogens

Some work has also been done on the use of the halogens, particularly bromine, as depolarizers [P 121,142]. It has been reported [A 19] that a high exchange current is involved for the Br_2/Br^- couple, so that good depolarizing characteristics should be expected; the energy density of Zn/Br_2 does approach that of Zn/AgO.

The reactions involved in the discharge of bromine depend on pH. The theoretical potential for the Br_2/Br^- couple is $+1.087$ V (*versus* the normal hydrogen electrode). However, in alkaline solution, the following reaction takes place [A 8]:

$$3\,Br_2 + 6\,OH^- \rightarrow 5\,Br^- + BrO_3^- + 3\,H_2O \qquad \varDelta G = -6.2\,kcal$$

BrO_3^- will then discharge as follows:

$$BrO_3^- + 3\,H_2O + 6e^- \rightarrow Br^- + 6\,OH^- \qquad E^0 = +0.61\,V$$

Note that $\frac{1}{6}$ the capacity of bromide has been lost as has 0.48 V. However, because of the "leveling effect" of the caustic electrolyte discussed above, the potential of 1.087 V would not be available in the absence of the bromate reaction.

In acid system, the bromate-formation reaction is actually reversed:

$$5\,Br^- + BrO_3^- + 6\,H^+ \rightarrow 3\,Br_2 + 3\,H_2O \qquad \varDelta G = -5.2\,kcal$$

so that the theoretical potential should be approached.

More of a problem is the tendency to form Br_3^- ions, which diffuse to the negative and are chemically short-circuited. This is particularly serious with the $MgBr_2$ electrolytes often used in magnesium anode cells. One solution to this problem involves absorption of the bromine on activated carbon [P 28]. Another is based on surrounding the positive with an alkyl ammonium halide to hold the free halogen. For example, one molecule of tetramethyl ammonium bromide is capable of binding nine atoms of bromine in a nonionic bond and releasing them on discharge. Supposedly, the bromine may be replaced by chlorine or iodine. A separator is also used to retard transfer [P 27].

The chloride–chlorine couple

$$Cl_2 + 2e^- = 2\,Cl^- \qquad E^0 = +1.3595\ V$$

is reversible, and chlorine is generally a more effective oxidizing agent than oxygen. It is to be expected, from the high potential, that water would be rapidly oxidized at a chlorine electrode. That this process does not occur at a fast rate is due to the high overvoltage associated with the deposition of oxygen. The Zn/Cl_2 system has a theoretical open circuit voltage of 2.12 V; in practice, 2.02 V has been observed [P 190]; the theoretical capacity Q_0 is 380 W-hr/lb. Some attempt has been made to apply this system to practical devices [P 191].

The general problems involved in implementing this cell may be considered typical of the use of gaseous depolarizers in battery systems. A primary reserve battery will be considered as an example. It is first necessary to get reactant gas to the cathode–electrolyte interface. Dual-porosity carbon cathodes have been described. The fine pores contain the electrolyte, while the large pores are intended to entrap and diffuse the gas. The term "converter–depolarizer" has been applied to this element. An aqueous $ZnCl_2$ electrolyte is immobilized in a porous membrane. The negative consists of a nonporous zinc sheet [P 192]. It is possible to construct these electrodes so that the entire cell is not more than $\frac{1}{4}$ in. thick. An electrically conductive, moisture-impervious coating is placed between the dry cathode and the anode of the next cell. For long-term stability in the nonactivated condition, the cell compartment should initially either be under vacuum [P 19] or contain an inert gas, e.g., nitrogen, rather than air [P 194]. To prevent an excessive loss of moisture, the diffuser section of the cathode is waterproofed [P 195].

The next problem is that of bringing the chlorine to the battery stack. For minimal volume and weight, chlorine is best stored as a liquid [vapor pressure at 70°F is 85.3 psig and density (liquid) is 1.468 g/ml]. For rapid activation of the battery, heat is needed within the battery casing to effect a rapid change of liquid chlorine into the gaseous form. This can be accomplished by bringing liquid chlorine into contact with a foil or filament of a metal having a high heat of reaction with chlorine, for example, aluminum, magnesium, tin, or zinc. The foil is coated with an adhesive which preferably also reacts with the chlorine to produce heat [P 193].

A design has been suggested [P 196] for the application of this system to electric-powered submarine torpedoes. The liquid chlorine is stored within the package and vaporized into the cell through a valve structure. The cells are kept mechanically under pressure [P 197] to maintain a constant anode–cathode separation as the zinc is consumed.

A Zn/Cl_2 system has also been investigated for possible application as a battery for guided missiles [A 5]. The cells were of the bipolar type with the gas diffusion cathode of one cell bonded to the metal anode of the next cell in series. The electrolyte (23.6% $CaCl_2$ and 7.5% $ZnCl_2$), immobilized in a siliceous paste, is installed at the time of manufacture. Activation consists of allowing chlorine to vaporize into the pressure container holding the battery. The device had to be used within 15 min after the chlorine had been admitted. The experimental problems were concerned principally with maintaining an adequate shelf life. A number of difficulties existed, e.g., successful bonding of the anode to the cathode and loss of moisture from the cell. Most serious, however, was the evolution of hydrogen due to corrosion of the zinc. Besides the loss of active material, this resulted in an increase in pressure which could eventually destroy the battery. A wet shelf life of one year was obtained [A 194].

These cells were intended to operate at high discharge rates, e.g., times as short as 1 min. Early tests showed capacities of 80 A-min/cell with peak voltages of 1.5–1.6 V/cell at 0.75 A/in². The specific energy was 17 W-hr/lb excluding the weight of the casing; the energy–volume density was 1.2 W-hr/in³. A material balance indicated a zinc use efficiency of 63%.

The most reactive of the halogens is fluorine:

$$F_2 + 2e^- = 2\,F^- \qquad\qquad E^0 = 2.87\ V$$

A Zn/F_2 cell would have a theoretical capacity of 868 W-hr/lb, based on the above voltage for the F_2/F^- couple. If a half-cell voltage of 1.36 V were used in the calculations (which is more compatible with aqueous systems), Q_0 becomes 510 W-hr/lb. A Li/F_2 cell has a theoretical open circuit voltage of 5.9 V and a Q_0 of 2780 W-hr/lb, which may well represent the ultimate in chemical batteries.

As will be discussed in the section dealing with anhydrous HF, the fluorine electrode has generally been observed to be irreversible. At best, the (H_2/HF, KF/F_2, Pt) cell had an open circuit potential of 2.768 V at 0°C. Fluorine reacts with water, giving oxygen mixed with ozone, hydrogen peroxide, and a number of other oxidizing substances [A 6]. Although this represents a loss in efficiency, it is to be expected that these reaction products would be electrochemically active.

A number of formidable materials and handling problems exist because of the highly reactive nature of fluorine. Nevertheless, the use of this material offers promise of a high-energy system, even if only a small fraction of the available energy can be achieved experimentally. Some work has been reported recently on the use of HF and CsF electrolytes, in conjunction with the electrochemical oxidation of hydrocarbons at 110°C [A 7]. The cell plates were constructed of Teflon; the electrodes were platinum black and carbon.

Oxy-Halogen Salts

Bromate is another of the common oxidizing agents which could be employed as a depolarizer. BrO_3^- is a faster oxidizing agent than ClO_3^- and in concentrated acid it will oxidize water to oxygen. The acid reaction is complicated by the following:

$$5 \text{ Br}^- + \text{BrO}_3^- + 6 \text{ H}^+ \rightarrow 3 \text{ Br}_2 + 3 \text{ H}_2\text{O}$$

The problems attendant on the use of bromine have been discussed. In base, however, the reaction is reversed, so that a cathode couple characteristic of bromate should be observed. The alkali salts of this material are soluble, so that, again, it would best be used in a primary reserve cell. The Q_0 for $Zn/LiBrO_3$ is 356 W-hr/lb, a result of its low equivalent weight and favorable potential. The mercuric bromate $Hg(BrO_3)_2$ is only slightly soluble in water (0.81 g/liter at 25°C), so that it might be possible to employ this material as a depolarizer in a dry cell. Q_0 for $Zn/Hg(BrO_3)_2$ is 293 W-hr/lb.

A similar preliminary evaluation can be made of iodate, also a strong oxidizing agent. In acid solution, iodate will oxidize iodide to iodine. This reaction is dependent on pH, however, so that in basic solution the end product is iodide. The system $Zn/KOH/LiIO_3$ has a Q_0 of 260 W-hr/lb. The calcium salt is somewhat less soluble and, hence, more suitable for dry-cell applications. Q_0 for $Zn/Ca(IO_3)_2$ is 272 W-hr/lb.

A primary reserve battery has been developed comprising a zinc anode, a potassium iodate cathodic reactant, and an electrolyte consisting substantially of sulfuric acid, *i.e.*, 8 N H_2SO_4 + 0.5 N HCl + 2% $HgCl_2$ [A 292]. It was found that mercuric chloride was very effective in reducing the formation of hydrogen, while the 0.5 N HCl increased the voltage of the battery at low temperature (0°C). The discharge reactions are probably as follows:

$$Zn^{+2} + 2e^- \rightarrow Zn \qquad E^0 = -0.763 \text{ V}$$
$$IO_3^- + 6 H^+ + 5e^- \rightarrow \tfrac{1}{2} I_2 + 3 H_2O \qquad E^0 = +1.195 \text{ V}$$

The positive was formed from a blend of KIO_3 (57.1%), graphite (40.8%), and acetylene black (2.1%). This mix was bonded, with polyvinyl acetate, to a silver-clad zinc plate. In the cell configuration studied, the cathode current efficiency was approximately 40–50%.

A reserve battery of this type was capable of reasonable current drain rates, typically 0.092 A/cm² (0.6 A/in.²), at a discharge potential of 1.6 V from 7 min. A typical self-contained battery, capable of operating at the above performance level, had an energy output of 2.32 W-min/cm³ and 14.2 W-hr/lb. Procedures have been developed for manufacturing low-cost cells in a simplified type of mechanical battery construction; prototype batteries have been constructed. Similar cells were constructed with bromate ion as the positive. These cells yielded voltage–time curves that were considerably less flat than those for cells using the iodate.

The periodates are somewhat stronger oxidizing agents. For example, an acid solution of IO_4^- will oxidize manganous ion to permanganate. It would be expected that, because of the high potential (+1.6 V), acid solutions of periodate would oxidize water to oxygen. The same should hold true in basic solution where the cathode reaction is as follows:

$$H_3IO_6^{-2} + 2e^- \rightarrow IO_3^- + 3 OH^- \qquad E^0 = +0.7 \text{ V}$$

It would be expected that IO_3^- would then be reduced electrochemically to I^-. The Q_0 for the $Zn/KOH/LiIO_4$ system is 310 W-hr/lb. This value was computed with the capacity of periodate, i.e., an 8-electron change, and the potential of IO_3^-, i.e., +0.26 V.

Cells of the following type have also been constructed:

$$Zn/H_2SO_4(dil.)/NaClO_3, C$$

The acid concentration is adjusted to avoid autodecomposition of $NaClO_3$ [P 405].

Halides

Silver chloride and cuprous chloride have been used as depolarizers in conjunction with seawater-activated batteries for powering torpedoes [P 122] and signal buoys. The capacity of the CuCl system is somewhat greater and, of course, it is less expensive.

AgCl electrodes are conveniently formed by chemically or electro-chemically depositing a coating onto silver. If the plate is built up chemically, the outer layers are not in electrical contact with the conducting substrate. Thus, the outer layers are not available for discharge and the reaction front must move inside out. The total capacity is not limited as much as the rate of discharge. To improve on this, it has been suggested that filaments of silver be formed throughout the AgCl deposit. This can be achieved by reducing the outer walls of porous AgCl with NH_2OH. Thus, the pore walls serve as conductive silver filaments passing through the mass [P 123]. It has also been suggested that, after AgCl is deposited, the plates be rolled, thus orientating the halide crystal parallel to the surface to improve contact with the substrate [P 124]. Apparently, the more conventional technique for providing reduced silver within the plate is immersion in a photographic film developer [P 404].

Besides problems of electrical conductivity, the use of CuCl cathodes is limited by the poor mechanical stability of the material. The inclusion of AgCl in the mix has been reported to improve the stability [P 125] of the positive somewhat. It has also been suggested that the cathode mix be plastic-bonded to a conductive plate rather than be formed by simple compression with carbon [P 127-129]. As a result, the rate of dissolution is decreased and the cell can be discharged over a period of hours. An effective support material is porous copper [P 120], since CuCl can be conveniently formed within the pores of the plaque by the following reaction:

$$Cu + CuCl_2 \rightarrow 2\ CuCl$$

The theoretical capacities of cells employing either CuCl or AgCl are low, and, hence, the applications would be expected to be of the "special" variety where weight is not a problem.

The addition of elemental sulfur to the cuprous chloride active material increases the electrode potential throughout discharge by approximately 0.4 V when operated in neutral or acidic media. The greatest benefit is attained by the addition of approximately stoichiometric quantities of sulfur. Cupric sulfide, rather than metallic copper, is the final discharge product [A 212]. The reaction responsible for this potential increase is

$$e^- + CuCl + CuS \rightarrow Cu_2S + Cl^-$$

where CuS is formed as follows:

$$2\,CuCl + S \rightarrow CuS + Cu^{+2} + 2Cl^-$$

Very little information is available on the discharge of fluoride compounds in aqueous solutions. It has been reported that cesium tetrafluoride, titanium tetrafluoride, and potassium–titanium fluoride are suitable in aqueous fluoride solutions [P 422]. A cell having a titanium negative, an NH_4F electrolyte, and a positive of silver fluoride (obtained by the reaction of silver nitrate and NH_4F) had an open circuit potential of 1.75–1.8 V [P 422].

Dichromates

Dichromates have also been used as depolarizers (*e.g.*, [P 115,143]). The equivalent weights in acid and base are low. However, the rate of the $Cr_2O_7^{-2}/Cr^{+3}$ reaction in 1 M acid is slow and the potential of the CrO_4^- in base is low. The Q_0 for the Zn/Li_2CrO_4 system is only 165 W-hr/lb. Furthermore, the reduction in alkaline solution is not complete; a basic chromic chromate, $CrOHCrO_4$, or some other intermediate is formed.

Bismuthates

Another inorganic oxidizer is sodium bismuthate [P 116]. The potential is high, *e.g.*, +0.56 V in base. However, the equivalent weight is also high. A couple of $Zn/LiBiO_3$ has a theoretical capacity of 135 W-hr/lb. It has also been suggested [P 117] that Bi_2O_3 be used as an oxidant. Not only is the equivalent weight high, but also the potential is unfavorable.

Persulfates

In the presence of a silver catalyst, persulfate is a strong oxidant and, hence, should function as a depolarizer. A patent [P 113] discusses the use of this material in a water-activated, primary reserve battery, where long-term wet stability is not needed. A suitable cathode structure was formed by binding a mix of silver (or carbon) and persulfate with polyvinyl acetate to a silver sheet. In effect, persulfates can be considered an acid–peroxide mixture, the instability of the compound increasing with increasing pH. The primary disadvantages of this material are its high weight and low potential.

Nitrates

Concentrated nitric acid is known to be a strong oxidant with the production of NO. However, in low hydrogen-ion concentration, it is not only weak, but also very slow in its reaction rate. Nitrate may be reduced with copper or silver cathodes. Because of the more rapid reduction of nitrite, it is difficult to stop the nitrate reduction at the nitrite stage, NO being obtained. A theoretical capacity of 215 W-hr/lb is obtained for $Zn/LiNO_3$. The $Zn/LiNO_2$ system has a theoretical capacity of 108 W-hr/lb, due primarily to the poor potential associated with the nitrite/NO couple.

Permanganates

These compounds have been employed to some extent as inorganic depolarizers for aqueous battery systems [P 114,168]. MnO_4^- is, of course, a strong oxidizing agent, being reduced to Mn^{+2} in acid solution and to MnO_2 in basic solution. A cell composed of Zn/KOH (aq.)$/KMnO_4$, C should show an open circuit potential of 1.82 V and a Q_0 value of 205 W-hr/lb. The use of the lithium permanganate would increase this to 230 W-hr/lb.

The shelf life of MnO_4^-, in the presence of electrolyte, may be limited, since there is a tendency to decompose slowly to MnO_2. Furthermore, the presence of MnO_2 in the electrolyte catalyzes the decomposition of permanganate. Both these effects should be of minor importance with primary reserve batteries, as should problems due to solubility in the electrolyte. It is envisioned that the battery would be operated only for a short period of time, e.g., 10 min.

Discharge curves for cells employing zinc anodes and cathodes of $KMnO_4$, $AgMnO_4$, $CuMnO_4$, $Ca(MnO_4)_2$, and $Ba(MnO_4)_2$ are given in Fig. 2-2 [P 357–359]. The cathode paste is formed by grinding together permanganate, a finely divided carbon (10 wt.%), and a water repellent which prevents loosening of the electrode. The electrolyte

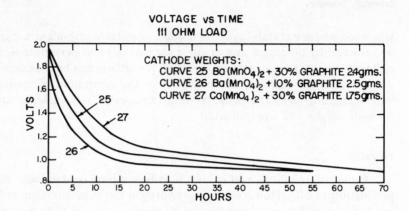

VOLTAGE vs TIME
111 OHM LOAD

CATHODE WEIGHTS:
CURVE 25 Ba(MnO₄)₂ + 30% GRAPHITE 24gms.
CURVE 26 Ba(MnO₄)₂ + 10% GRAPHITE 2.5gms.
CURVE 27 Ca(MnO₄)₂ + 30% GRAPHITE 1.75gms.

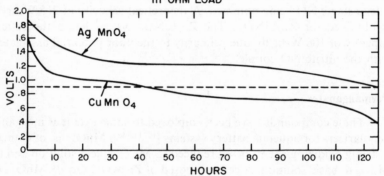

VOLTAGE vs TIME
111 OHM LOAD

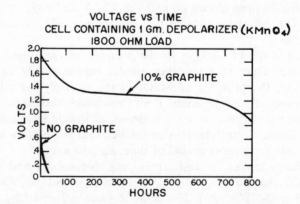

VOLTAGE vs TIME
CELL CONTAINING 1 Gm. DEPOLARIZER (KMnO₄)
1800 OHM LOAD

Fig. 2-2. Discharge curves for cells employing zinc anodes and cathodes of $KMnO_4$, $AgMnO_4$, $CuMnO_4$, $Ca(MnO_4)_2$, and $Ba(MnO_4)_2$ [P 357–359].

in permanganate cells is immobilized by a magnesium or aluminum silicate gel. The initial drop in voltage on load is due to a high-resistance, brown film formed early in the discharge of the cell. The voltage can be minimized by including in the $KMnO_4^-$ graphite mix a small amount of electrolyte absorbent which presumably maintains an electrolyte path [P 360].

Sulfur

Alkaline solutions of sodium sulfide containing dissolved sulfur have been found [A 314] to give rise to reproducible equilibrium potentials (-0.530 ± 5 mV) at inert electrodes. The potentials were unaffected by stirring, and the equilibrium rapidly recovered after it had been displaced by anodic or cathodic polarization. Furthermore, the equilibrium potential was almost independent of the nature of the electrode material. The exchange currents were determined to be of the order of 5×10^{-6} A/cm².

The cathodic discharge of sulfur apparently proceeds to the formation of polysulfides of the type S_4^{-2}. Since the material is ionic, it can diffuse from the cathode into the electrolyte, resulting in a loss of active material.

Recent work on the Mg/KSCN/S, C/Ag system employed an ion-exchange membrane to separate components, which increased coulombic efficiency from 7.8 to 17% [A 103].

The addition of elemental sulfur to copper oxide and to cuprous chloride has improved the discharge properties of these materials (see above). SO_2 has also been used as a depolarizing agent for copper, iron, and aluminum electrodes [A 268].

Carbides, Nitrides, and Borides

An examination of standard electrochemical reduction tables indicates that carbon, nitrogen, and oxygen can also be *formally* considered as oxidizing agents, *e.g.*,

$$C + 4H^+ + 4e^- = CH_4(g) \qquad E^0 = +0.13 \text{ V}$$

The activities of carbides, nitrides, borides, and silicides were studied [A 196]. Calcium hexaboride is reportedly stable against chemical attack by acid and basic aqueous solutions, but is electrochemically active to the extent that it promotes rapid corrosion of zinc when coupled with it in acid solution [A 195]. It was not possible to identify the reaction products, and further work on this material was terminated [A 196]. The electrochemical behaviors of ferrosilicon, ferroboron, and iron carbide were also studied. As anodes, the electrochemically active component was the iron. As cathodes, air was reduced on their surfaces in preference to reduction of silicon, boron, or carbon [A 197]. The sample did contain acetylene black, which has shown a small amount of activity as an oxygen electrode [A 198].

Manganic Compounds

Although trivalent manganese compounds are unstable in water, the rate of decomposition of $MnPO_4$ in acid is relatively slow. Dry cells of the following type have been built [A 197]:

$$Zn/NH_4Cl, ZnCl_2/NH_4Cl/MnPO_4 , C$$

The open circuit potential was of the order of 2 V; the cell potential was 1.6 V at the 100-mA rate.

The following cell was built to avoid the problem of electrolyte mixing:

$$Pb/H_2SO_4 (30\%)/MnPO_4 , C \qquad V_{oc} = 1.4 V$$

An average voltage of 1.0 V was obtained at the 100-mA rate. Greater insolubility was observed with an electrolyte of HBF_4 and cells of the type $Pb/HBF_4/MnPO_4$ were built [P 294]. Apparently, a valence change of 1 is observed for this compound, which gives a theoretical equivalent weight of 150.

Redox Resins [A 197]

The chemical and electrochemical behavior of resinous hydroquinone derivatives has also been studied. These materials were prepared by acid- or base-catalyzed reactions of hydroquinone with formaldehyde or with both phenol and formaldehyde. The base-catalyzed reaction products were acid-insoluble resins which could be oxidized and reduced chemically. However, they did not appear to be reduced easily by electrochemical means. Similar results were obtained with the acid-catalyzed reaction product.

This approach has been applied more recently to cells of the following type [P 394]:

$$Li/LiCl, DMF/polyquinone, Pt$$

These cells have had high internal resistance and have shown low efficiencies in the use of the positive material.

Hydroxylamine

Hydroxylamine has been reported superior to MnO_2 as a depolarizer in alkaline, $MgBr_2$, and ZnO_2^{-2} electrolytes [P 120]. The theoretical potential data indicates that in acid hydroxylamine is a strong oxidizing agent ($E^0 = +1.35$ V). However, the actual rate of reduction appears to be slow. Hydroxylamine is unstable with respect to decomposition into ammonium ions and nitrous oxide:

$$4NH_3OH^+ \rightarrow N_2O + 2NH_4^+ + 3H_2O + 2H^+$$

and the reaction is favored by alkali as the equation indicates.

Cupric Oxide

Cupric oxide, which has an equivalent weight of 79.5, has been used as a depolarizer, particularly at higher temperatures [P 361]. Zn/CuO has a theoretical capacity of 126 W-hr/lb. This battery is an old system, having been developed in 1882, and it has found considerable application in railway signalling devices.

The observed equilibrium potential coincides with the potential of the Cu/Cu_2O electrode, cuprous oxide being formed from cupric oxide and copper. The extent of this reaction depends on the electrical conductivity of the cuprous intermediate layer, which has semiconductor properties. Introduction of sulfur, an electron acceptor, into the active material has resulted in a manifold increase in the conductivity of hydrated cuprous oxide. This then accelerates renewal of the active hydrated cuprous oxide layer which is expended during discharge. The observed result is a 25% increase in cell voltage; e.g., 0.84–0.75 V, rather than 0.68–0.50 V, is observed at 4 mA/cm² [A 212].

Silver Oxides

Of the existing primary and secondary battery systems, those employing silver oxide positives most closely approach the criteria for high-energy batteries. This material is used in both primary and secondary cells.

The standard potential of the Ag/Ag_2O couple, measured electrochemically in alkaline solutions, is $+0.342$ V [A 36] and the potential of the AgO couple is $+0.599$ V [A 37]. Based on AgO, the theoretical capacity of silver oxide is 0.438 A-hr/g AgO. In practice, approximately 0.244 A-hr/g AgO is generally achieved, yielding an overall efficiency of 55.5% based on silver.

AgO has a conductivity of the order of 7×10^{-2} (Ω-cm)$^{-1}$, which is substantially higher than that of Ag_2O [10^{-8} (Ω-cm)$^{-1}$]. The conductivity increases with temperature [A 41], as is expected for semiconductors. The structure of AgO is a matter of some dispute. There appear to be at least three different modifications—monoclinic AgO, which is discussed below [A 345,346]; a tetragonal AgO, which appears to be a peroxide containing monovalent and divalent silver [A 347]; and a cubic AgO, which has the sodium chloride structure and which is probably a peroxide also [A 348]. Only monoclinic AgO has been definitely found in the oxidation of silver in alkaline solutions.

Monoclinic AgO is diamagnetic [A 41], a fact which is explained by assuming that monovalent and trivalent silver ions exist within the oxide lattice. This is supported by crystallographic studies, including X-ray [A 42] and neutron diffraction [A 36] studies, which show that the silver atoms in the AgO lattice are not equivalent. There are two different Ag–O distances, namely, 2.18 Å and 2.03 Å, corresponding to monovalent and trivalent silver [A 36]. The specific nature of AgO is still under study [A 303]; an extensive review of the subject is available [A 312].

Solid AgO is thermodynamically unstable in contact with alkaline solutions, its potential being positive to that for oxygen evolution. In spite of its thermodynamic instability, AgO decomposes very slowly in alkaline solutions, presumably because of a high oxygen overvoltage [A 42] and because the decomposition product, i.e., Ag_2O, may form a protective surface layer.

Another process for decomposition of AgO is direct reaction with the cellulose separators generally used in secondary cells. Therefore, it is necessary to prevent contact of electrode and separator. Various "inert" separator membranes have been explored. Another method is to coat the silver electrode with a suspension of magnesium hydroxide which, on air-drying, is converted into MgO and $MgCO_3$. In one comparative test, cycle life was increased in this way from 8–10 to

16–20 cycles [P 319]. The effect of the MgO coating on the internal
resistance is reportedly minor [P 319].

Discharge of a AgO plate at low rates takes place in two stages,
as shown in Fig. 2-3. Since these reactions are approximately 100%
efficient [A 38] (*i.e.*, all Ag_2O is discharged to silver), the curve indicates
that a considerable amount of AgO is formed on electrical charging.
At high rates, the two plateaus blend into one.

The existence of these two plateaus can cause complications
when close control of the cell voltage is required. This high-voltage
plateau can be eliminated by controlled charging, at the expense
of overall capacity, however. In the case of reserve batteries, the plateau
can be eliminated by thermally decomposing the peroxide to oxide at
about 100°C [P 317]. Chloride ion is reported to eliminate the high-
voltage plateau without any large loss in capacity [P 91,318]. Thin silver
electrodes formed by treating AgCl foil with alkali are mechanically
stable if the reaction is carried out after the plate is assembled, so
that the pressure of the membrane prevents disintegration of the silver
oxide plate. Such electrodes show only one charge and one discharge
plateau, but have the same capacity (A-hr/g Ag_2O) as conventional
plates [A 318].

It has been shown [P 101] that the utilization of silver at the
argentous level is improved by the addition of small amounts (~1%)
of lead to the powder. Apparently, 1% Pd is also effective [A 213]. For

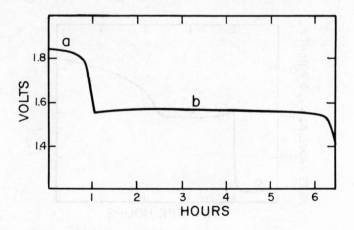

Fig. 2-3. Discharge curve for silver–zinc alkali cell.   (a) Reduction of AgO [A 34].
(b) Reduction of Ag_2O.

example, it has been shown [A 209] that the utilization of silver at the argentous level is approximately 71% in the fifteenth cycle for hot-pressed plates formed from 0.5μ silver and containing 1% Pb or 1% Pd. The discharge current was 120 mA/in². Apparently, the palladium also accelerates the uptake of hydrogen by the silver plate.

The addition of sulfide ion to the silver electrode may also increase capacity. For example, it has been reported that *metalloceramic* silver plates treated with sulfide ion had 10–13% greater capacity than untreated electrodes and that the upper state discharge curve was prolonged [A 214,215]. A similar effect was noted for Li_2O. The oxygen overvoltage was also increased. HgO had no noticeable effect. It is argued that Li_2O and sulfur affect the semiconductor properties of Ag_2O, facilitating a more complete oxidation of metallic silver [A 215].

The anodic charging of silver in an alkaline electrolyte is also a two-step process, as shown in Fig. 2-4. When the layer of Ag_2O first produced on the surface of the silver has reached a critical thickness, mass transport through it becomes slow and Ag_2O is oxidized to AgO (plateau b in Fig. 2-4). Further anodization leads to rapid evolution of oxygen and to further oxidation of metallic silver at a low rate. There is some evidence for the formation of an extremely unstable "higher oxide" during prolonged anodization, particularly at high current densities [A 303].

Measurements of the double-layer capacity and of the interfacial resistance during the anodic formation of Ag_2O films [A 35] give some

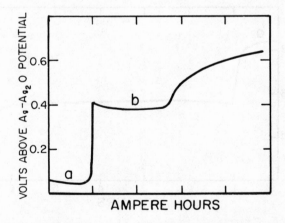

Fig. 2-4. Charge curve of silver electrode in alkaline solution.   (a) Formation of
Ag_2O.   (b) Formation of AgO [A 34].

insight into the specific mechanism of oxidation. The double-layer capacity decreases during building of the Ag_2O layer. At the same time, the interfacial resistance increases sharply, indicating the buildup of an insulating or passivating surface film.

This film of Ag_2O inhibits the migration of OH^- into the crystal and it is also electrically insulating, having a specific conductivity of about 10^{-8} $(\Omega\text{-cm})^{-1}$ [A 36]. These properties diminish greatly the rate of charge acceptance [A 39,40]. Thus, it has been shown that low rates of charge (12–24-hr rate) are necessary for good performance. Apparently, the slow step is associated with migration of OH^- through the film. However, Zn/AgO batteries can be discharged at high rates; this may be because discharge of the peroxide does not yield a continuous Ag_2O layer.

Electrodes of silver peroxide can be formed by a variety of techniques other than electrochemical [P 356], e.g., from a wet paste [P 320], by compressing a dry powder [P 321], or by plastic-bonding with polystyrene [P 317] or polyethylene [P 323]. AgO is particularly useful as a positive for reserve cells for the following reasons: (1) its capacity is high; (2) the electrode is spongy, and the pores tend to enlarge as AgO is reduced to silver; (3) the electrode is mechanically strong and does not deteriorate in physical strength when reduction to metallic silver occurs; (4) when employed as a primary, it can deliver full power within 1 min after activation with electrolyte.

Manganese Dioxide

This material has been used extensively as a depolarizer in the LeClanché cells. The equivalent weight is apparently a function of pH. For example, manganese dioxide electrodeposited on spectroscopic graphite rods was discharged in 0.1–9.0 M KOH at constant current in the range 11–300 mA/g of MnO_2. "The discharge curves showed two steps: the first step from $MnO_{2.0}$ to $MnO_{1.5}$ and the second step from $MnO_{1.5}$ to $MnO_{1.0}$. The shape of the discharge curve of the first step was not strongly influenced by the discharge conditions, such as current density, KOH concentration, thickness of MnO_2 layer, and addition of complexing agent (triethanolamine) to the electrolyte; the second step was strongly influenced by the discharge conditions, and the potential was fairly constant during the second step at low current densities and high KOH concentration. In order to explain the constant voltage discharge curve, a mechanism is proposed based

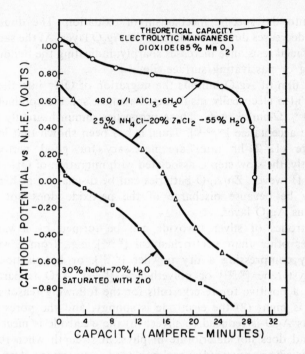

Fig. 2-5. Effect of electrolyte on the performance characteristics of an electrolytic MnO$_2$ cathode material [A 50].

on the solubilities of Mn(II) and Mn(III) ions in KOH electrolyte" [A 304].

The performance of MnO$_2$ in an acid electrolyte, *i.e.*, AlCl$_3$ solution [A 50], is illustrated by the coulometric data in Fig. 2-5 for an electrolytically prepared manganese dioxide (85% MnO$_2$). In the acidic aluminum chloride electrolyte, MnO$_2$ operates at a potential 0.4–0.5 V higher than in the NH$_4$Cl–ZnCl$_2$ electrolyte and gives twice the ampere-minute capacity at an end voltage of +0.25 V. This capacity corresponds to a theoretical two-electron change per molecule of MnO$_2$, indicating that reduction probably takes place in accordance with the following equation [A 50]:

$$MnO_2 + 4H^+ + 2e^- \rightarrow Mn^{+2} + 2H_2O$$

Under these acidic conditions, a Zn/MnO$_2$ couple has a theoretical capacity of 280 W-hr/lb. An open circuit potential of 1.75 V was assumed, as well as a two-electron change for MnO_2. Combination of MnO_2 with an anode of lower equivalent weight, e.g., magnesium or titanium (and at the same potential) yields a Q_0 value of 395 W-hr/lb.

Nickel Oxide

The equivalent weight of NiO is 38 for the theoretical reaction:

$$NiO + H_2O + 2e^- \rightarrow Ni + 2OH^-$$

It has been established that: (1) in practice, the actual discharge involves one electron, and (2) nickel metal is not produced. In spite of many studies, the characterization of nickel oxides, the identification of the reactions which take place on charging and discharging, and the elucidation of aging processes during storage still prove to be difficult and controversial problems.

A variety of exotic compounds of nickel, oxygen, and hydrogen has been prepared [A 217]; however, the actual compounds which were claimed to be due to charge and discharge of a nickel "oxide" electrode are as follows: β-NiOOH; $Ni_2O_3 \cdot 2H_2O$; $Ni_2O_3 \cdot H_2O$; $Ni(OH)_2$; and $Ni_3O_2(OH)_4$. X-ray diffraction measurements indicated that β-NiOOH and $Ni(OH)_2$ exist under ordinary charge–discharge conditions. Under extreme overcharge conditions, $Ni_2O_3 \cdot H_2O$ may also be formed. The existence of higher oxides helps explain an excess capacity of 10–15% found upon discharge of freshly charged plates [A 218]. This excess has also been attributed to a monolayer surface phase of adsorbed oxygen-containing radicals, such as O^{-2} or OH^-, on the conducting metal surface [A 219]. Apparently, this excess capacity will self-discharge within 48 hr on wet stand [A 308].

The structures and composition of the lower-valence oxide and hydroxide are relatively well-established. $Ni(OH)_2$ is a layer structure in which every nickel atom is surrounded by six hydroxyl groups. Every hydroxyl group forms three bonds to nickel atoms in its own layer and, at the same time, is in contact with three hydroxyls of the adjacent layer. $Ni(OH)_2$ generally contains varying amounts of water up to $Ni(OH)_2 \cdot 1.5H_2O$.

The infrared spectrum of sintered nickel oxide electrode, determined in the region 4000–250 cm^{-1}, confirms many of these conclusions. "The spectral data show the discharged state to be $Ni(OH)_2$ having a hexagonal layer structure isomorphic with the space group $D_{3d}{}^3$. The hydroxyl groups are parallel to the c-axis of the crystal and are free in the sense that hydrogen bonding is absent. Associated with this structure is a relatively small amount of water which is trapped in the crystal lattice through the formation of coordinate-covalent bonds with the nickel ions" [A 305].

The charging process is accompanied by the formation of a hydrogen-bonded structure possessing a higher degree of crystal symmetry than is found in the discharged state. During discharge, these hydrogen bonds are continuously broken with the formation of free hydroxyl groups.

The active material of the nickel positive is apparently β-NiOOH (e.g., [A 220]). However, chemical analysis of active oxygen in charged plates show oxygen–nickel ratios above that for NiOOH. This was attributed to the presence of some tetravalent nickel ions. The extent of hydration of the active materials of charged and discharged electrodes is a matter of controversy. Apparently, some water of hydration is necessary for the satisfactory performance of this electrode. In fact, a major change on storage is the aging of $Ni(OH)_2$ with the loss of H_2O and conversion of amorphous $Ni(OH)_2$ into a more or less crystalline form [A 221]. This water is apparently lost from between the layers of $Ni(OH)_2$ crystallites. Besides recrystallization effects, the charged nickel oxide also self-discharges through oxygen evolution (e.g., [A 222]).

Thus, the exact equation for charging and discharging nickel "oxide" is quite complex. Nevertheless, the following three conclusions can be made: (1) the discharge reaction involves one electron per nickel atom; (2) the performance is intimately tied up with water and hydroxide-ion transfer into and out of the plate, and, thus, the charge and discharge of this electrode in a nonaqueous system would have to take place by a drastically different mechanism; and (3) the equivalent weight of the active, charged material is at least 92.

The incorporation of lithium into the nickel oxide plate significantly improves the discharge characteristics. This effect has been generally attributed to a decrease in ohmic resistance of the plate material. It has also been shown by X-ray diffraction studies that the presence of LiOH in the KOH electrolyte also gives rise to a more finely divided active material [A 279].

Cobalt Oxide

It has been reported that cobalt metal can be oxidized anodically in alkaline electrolyte to cobalt oxides [P 425]. For the highest oxide of cobalt, a potential of $+0.7$ V against a hydrogen electrode has been measured [P 425]. Electrodes of cobalt oxide show reversibility, can store a high amount of oxygen per unit volume, and are twice as light as electrodes of silver oxide. The replacement of the silver oxide electrode in a silver–zinc alkaline battery system would be attractive from a cost standpoint. At the present market price, cobalt can be purchased at approximately one-fifth the price of silver.

The open circuit voltage of a cobalt–magnesium cell approaches the theoretical cell voltage value of 3 V; a cobalt–zinc cell had an output voltage of 2 V. A quaternary ammonium hydroxide electrolyte was used for these cells [P 425]; the cobalt oxide electrode was formed by precipitation from cobalt nitrate with caustic solution.

However, the discharge capacity does not appear to be high. For example, the cobalt hydroxide electrode has been studied in some detail [A 318,319] with an electrolyte of 1 M KOH. The starting material, $Co(OH)_2$, was formed electrochemically. Under the experimental conditions employed, $Co(OH)_2$ was readily oxidized to CoOOH at a potential of approximately $+0.13$ V (versus Hg/HgO); the reverse reaction appeared to be rather slow, unlike the reduction of NiOOH. Apparently, a small amount of higher oxide can be formed, corresponding to a composition of about $CoO_{1.05}$. The amount of material so formed constitutes at most a monolayer; the net capacity is, thus, rather low.

Silver Sulfide

Silver sulfide has an equivalent weight of 140, which is too high for use in a high-energy-density battery. However, Ag_2S has a solubility product of 1.6×10^{-49} compared to 1.7×10^{-10} for AgCl, so that self-discharge due to chemical dissolution would be less and a long shelf life would be expected.

The potential of a reversible electrode of this type, i.e., an electrode of the second kind, can be expressed in terms of the solution concentration of the ion to which the electrode is reversible, in this case silver.

$$E = E^0 + 0.059 \log [Ag^+]$$

The silver-ion concentration is determined by the solubility product for the reaction

$$Ag_2S \rightarrow 2Ag^+ + S^{-2}$$

i.e.,

$$K_{sp} = [Ag^+]^2[S^{-2}]$$

Thus, for unit activity sulfide ion, the equilibrium potential for the Ag_2S/Ag electrode is -0.69 V.

Although the Ag_2S/Ag appears to be irreversible, the observed potential did obey the Nernst equation in the following form [A 276,277]:

$$E = b + (RT/nF)\ln m$$

where $b = 0.691$ V, $n = 1.61$, and m is the sulfide content.

Such a system has been suggested [A 216] for use as a long-shelf-life, high-voltage, low-current source. A zinc electrode was coupled with a Ag_2S positive; the electrolyte was $ZnCl_2$ saturated with zincate. The sulfide ion, released to the electrolyte by the discharge reaction, formed ZnS. The solubility of this compound is 1.2×10^{-23}, so that in 8 M Zn^{+2} the sulfide-ion concentration is 1.5×10^{-24}. Thus, the silver-ion concentration is 3.3×10^{-13} and the half-cell potential of the silver electrode is $+0.06$ V.

It has also been assumed [A 216] that the complex-ion concentration is negligible. However, it can be shown that the AgS^- ion contributes 4×10^{-13} moles/liter of soluble silver. The soluble silver halide complexes must also be considered. With the chloride concentration used in this cell, approximately 10^{-10} moles/liter of soluble silver is generated, a factor of 10^2 more than is contributed by the sulfide or its complexes. Furthermore, sulfides are slow to come to equilibrium, which could lead to a slow discharge rate.

This is, nevertheless, an improvement over the AgCl/Ag couple in the same medium. Calculations of complex-ion formation indicate $\sim 10^{-3}$ moles/liter of soluble silver in 3.7 M Cl$^-$. A possible solution to this problem is to use a noncomplexing anion, such as ClO_4^-.

ORGANIC POSITIVES

The use of organic compounds as depolarizers has received a good deal of attention (*e.g.*, [A 286,287; P 416-418]). Recently, an extensive survey has been made of the subject [A 288].

The intense interest in these systems stems from the fact that nitro compounds, because of their low equivalent weight, have theoretical energy densities that are far higher than many common inorganic cathodes. For example, the theoretical energy density of the m-DNB/Mg couple is estimated to be 855 W-hr/lb of reactants, including the 8 moles of water that are consumed with every mole of m-DNB:

$$m\text{-}C_6H_4(NO_2)_2 + 6Mg + 8H_2O \rightarrow C_6H_4(NH_2)_2 + 6Mg(OH)_2$$
$$E^0 = 2.67 \text{ V}$$

The experimental operating potentials of the m-DNB/Mg cells at their present state of development are several tenths of a volt less than those of Zn/MnO$_2$ cells, due to inefficiencies at both the magnesium and m-DNB electrodes.

Results of testing about 200 different organic compounds for depolarizer activity have been tabulated [A 288] with experimental and theoretical coulombic capacities and average operating potential drop from open circuit to a fixed load. The conclusions of this work are given below.

"The best organic depolarizers are generally one of four types: nitro compounds (RNO$_2$), positive halogen compounds (ROX and RNX), halogen addition compounds (R$_3$NX$_2$), and peroxides (ROOR). The positive halogen compounds and the halogen addition compounds characteristically discharge at high potentials with high efficiency even at high drain rates. The peroxides have high reversible potentials, but generally are reduced with lower efficiencies. All except the nitro compounds have limited coulombic (and energy) capacity, since their reduction involves only a two-electron change per active group ($-OX$, $-NX$, $-O-O-$, $-NX_2$). Most of the positive halogen compounds have limited stability in aqueous media.

"Nitro compounds that have average operating potentials, energy capacities, and coulombic efficiencies greater than those of m-DNB are listed as follows: 2, 5-dinitrobenzoic acid; 3, 6-dinitro-phthalic acid and its anhydride; 1, 4-dinitropyromellitic acid and its silver salt; o- and p-DNB; 4-nitropyridine; picric acid; 1, 4, 5-trinitro-naphthalene; 1, 3, 5-trinitrobenzene; and 2, 4, 5-trinitrotoluene.

"Compounds that have an energy capacity over 20 W-min/g when tested in a thin electrode configuration are the following positive halogen compounds: 2, 5-dinitropyrrole; 1. 4, 5-trinitronaphthalene; picric acid; and o- and p-DNB.

"A semiquantitative measure of the effect of substituents on the reduction potential of the active group ($-NO_2$) was found by correlating the initial reduction potential with the Hammett sigma constant. The correlation held only for the earliest part of the discharge. Cell characteristics for the longest part of the discharge period were determined by physical factors (*i.e.*, water retention of bobbin, reaction products, and pH).

"Meta-DNB cathodes were discharged much more efficiently in acid electrolyte (100% in 2M HBr · MgBr$_2$ at 300 mA/g) than in neutral electrolyte (10% in 0.85M MgBr$_2$ at 50 mA/g). The same cell in 2M Mg(ClO$_4$)$_2$ electrolyte had an efficiency of more than 17% at 50 mA/g. Active metal anodes, such as magnesium, are attacked by acid, so the acid electrolyte cannot be used in conventional m-DNB/Mg cells.

"Polarographic and spectrophotometric methods have been used to determine the solubility of m-DNB in various aqueous and mixed-solvent electrolytes. The effect of organic additives in the bobbin was found to be detrimental. Bobbin discharge characteristics and solubility of m-DNB in aqueous magnesium-ion electrolytes can be correlated, *e.g.*, m-DNB has a high solubility in concentrated Mg(ClO$_4$)$_2$, and this electrolyte is also excellent for the bobbin discharge.

"Mg(ClO$_4$)$_2$ was the best neutral electrolyte found for the m-DNB bobbin. It is believed that its advantage over MgBr$_2$, for example, is its salting-in effect on m-DNB. Mg(NO$_3$)$_2$ also salts in m-DNB, and a bobbin with this electrolyte is a better cathode than one with MgBr$_2$. The concentration of Mg(ClO$_4$)$_2$ should presumably be 2M or greater. The maximum conductivity of Mg(ClO$_4$)$_2$ is at 2M. However, the maximum solubility of m-DNB is in saturated Mg(ClO$_4$)$_2$ (about 3.3M). In 3.3M Mg(ClO$_4$)$_2$, complex soluble oxy-magnesium species form at pH 6.5 in preference to the insoluble Mg(OH)$_2$ that precipitates at pH 7.5 [for 2M Mg(ClO$_4$)$_2$]. Thus, if corrosion of magnesium is tolerable in saturated Mg(ClO$_4$)$_2$, this electrolyte could be best for a m-DNB/Mg battery.

"Electrolyte content is of primary importance to the discharge of a m-DNB battery bobbin. A high electrolyte retention is needed by the carbon. If a carbon has a low apparent density but does not absorb water readily, water retention can be improved by a vacuum impregnation of the electrolyte. The use of wetting agents also increased the water retention significantly when a simple addition of electrolyte to carbon black is employed.

"Various catalytic materials were added to the m-DNB bobbin to improve its performance. Of the materials tested (silver, palladium, and nickel), only nickel had activity. However, its effect on potential was not as great as its effect on extent of discharge. If acid electrolytes can be tolerated, Th(NO$_3$)$_4$ addition to Mg(ClO$_4$)$_2$ will improve the m-DNB discharge. Formation of small, pure m-DNB crystals by rapid recrystallization into water also improves bobbin performance" [A 283].

The presence of V_2O_5 in the cathode mix apparently improves cell performance. However, there is a "deleterious interaction between a magnesium anode and the V_2O_5 additives, but because of the slight solubility of V_2O_5 (0.8 g/100 ml H_2O), the deleterious effects of the undesired interactions do not become apparent until after several months of cell storage" [A 362].

"Electrolytes of strong bases caused marked decreases in the capacity of m-DNB bobbins. The decrease was not caused by ohmic polarization, but rather because high pH favors the chemical reaction of the nitroso and hydroxylamine intermediates to form azoxy compounds. Electrolytes with cations that yield insoluble bases [e.g., $Mg(OH)_2$] cause large increases in IR polarization in the bobbin.

"The halogen addition compounds have good cathodic discharge characteristics. A study of their stability was made. The most stable complexes of monomeric bases are complexes of the type $NHXX_2$. The addition compounds of strong bases, such as $RNHXX_2$, tetramethyl ammonium chloride perbromide, and quinolinium chloride perbromide, are very stable. For example, tetramethyl ammonium chloride perbromide was 100% stable in the dry state for 35 days at 55°C, and the compound retained 97% of its active halogen for 15 days in $MgBr_2$ electrolyte at 55°C. Simple halogen addition compounds of polymeric bases are stabilized, e.g.,

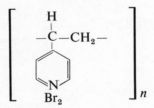

although the addition compounds of monomeric bases, e.g.,

are unstable.

"Simple mixtures of depolarizers with different characteristics discharge independently of each other when mixed together. No improvement in the discharge characteristics of either component of the following mixtures was observed: 1:1 m-DNB and p-nitrobenzoyl peroxide; 1:1 1:1 m-DNB and quinoline perbromide; and 1:1 1, 4, 5-trinitronaphthalene and 1, 4, 5, 8-tetranitronaphthalene.

"Nitro compounds with high solubility in electrolyte, such as tetrasodium dinitropyromellitate, potassium 2, 4-dinitrobenzene sulfonate, and disodium dinitrophthalate, effectively inhibited magnesium corrosion" [A 288].

Chapter 3

Electrochemically Active Materials: Nonaqueous, Inorganic Electrolyte Systems

LIQUID AMMONIA

Electrolyte Properties

The physical properties of liquid NH_3 are compared in Table 3-I to those of other solvents [A 134]. Liquid ammonia, like water, is a protonic solvent, dissociating into ammonium ions and amide ions:

$$2NH_3 \rightarrow NH_4^+ + NH_2^-$$

and, like water, it is not conductive because of a very small dissociation constant ($K = 10^{-33}$). The ammonium ion is the solvated proton, and, hence, ammonium salts, such as NH_4SCN, are strong acids in liquid ammonia. Hydrogen evolution, through corrosion, occurs on metal electrodes in liquid ammonia just as in water, but, this reaction occurs at more negative potentials, since the concentration of NH_4^+ is smaller in pure NH_3 (as low as 10^{-15}) compared with the concentration of H_3O^+ in pure $H_2O(10^{-7})$. Of course, the presence of acids, such as NH_4SCN, increases the corrosion rate.

The solubilities of some ionic salts in liquid NH_3 are listed in Table 3-II [A 99]. The change of conductivity with the concentration of one solute, KSCN, is shown in Fig. 3-1 [A 98]. Somewhat higher conductivities can be achieved with the acidic NH_4SCN, *e.g.*, 0.4 $(\Omega\text{-cm})^{-1}$ at room temperature [A 97,98].

A boiling-point elevation is obtained in some concentrated salt solutions. This effect is shown for KSCN in Fig. 3-2 [A 98]. A more

TABLE 3-I

Physical Properties of Solvents

	Water	Ammonia	Hydrazine	Pyridine	Sulfur dioxide
Formula	H_2O	NH_3	N_2H_4	C_5H_5N	SO_2
Specific conductivity $(\Omega\text{-cm})^{-1}$	4×10^{-8}	5×10^{-11}	2×10^{-6}	5×10^{-8}	4×10^{-8}
Cation	H_3O^+	NH_4^+	$N_2H_5^+$	$C_5H_6N^+$	SO^{+2}
Anion	OH^-	NH_2^-	$N_2H_3^-$	$C_5H_4N^-$	SO_3^{-2}
Melting point (°C)	0	−78	2	−40	−73
Boiling point (°C)	100	−33	114	116	−10
Dielectric constant	81 (18°C)	22 (−33°C)	52 (25°C)	12 (20°C)	18 (−20°C)
Dipole moment (D)	1.8	1.5	1.8	2.1	1.6
Viscosity (cP)	1.0 (20°C)	0.276 (−40°C)	0.974 (20°C)	0.945 (20°C)	0.428 (−10°C)
$\Lambda_0 KI$	149 (25°C)	332 (−33.5°C)	133 (25°C)	80 (25°C)	222 (−10°C)

TABLE 3-II

Solubilities in Liquid NH₃ at 25°C

Substance	Solubility (moles/mole NH₃)
NH₄SCN	0.861
NH₄NO₃	0.829
LiNO₃	0.602
NaSCN	0.432
NH₄I	0.432
NH₄Br	0.413
NH₄Cl	0.326
NH₄ acetate	0.332
Li	0.28
K	0.23
NaBr	0.228
NH₄ClO₄	0.200
NaNO₂	0.196
KI	0.186
RbF	0.172
NaI	0.161
(NH₄)₂S	0.139
CsI	0.100

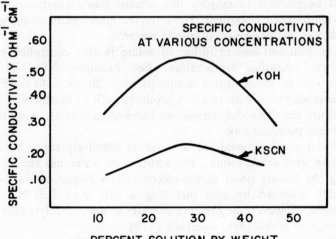

Fig. 3-1. Specific conductivity of KSCN–NH₃ and KOH–H₂O solutions [A 98].

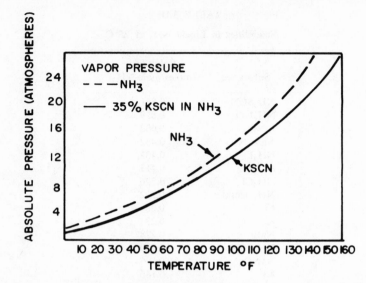

Fig. 3-2. Vapor pressure–temperature curves for liquid NH_3 and for a solution of KSCN–NH_3 [A 98].

pronounced change is observed with a solute of NH_4SCN. For example [A 104], a solution containing 75 mol.% NH_4SCN in anhydrous ammonia boils at $+64°C$, whereas pure ammonia boils at $-33°C$. Thus, it should be possible to employ this solvent–solute combination as a room-temperature electrolyte. However, the high acidity of this solvent complicates the choice of an anode material.

The practical concentration of solute is also controlled by the minimum operating temperature. For example, $NH_4SCN \cdot NH_3$ freezes out of concentrated solutions at -20 to $-40°C$. Similar problems may exist with reaction products such as $Mg(SCN)_2$. When this occurs, the electrodes can become blocked, increasing concentration and ohmic polarizations.

Water can be considered as a solute effectively supplying NH_4^+ [P 32]. As with other solutes, the addition of water has the effect of raising the boiling point of the electrolyte. A higher conductivity is generally achieved by also including a salt, e.g., LiSCN, in the electrolyte. However, the presence of water in the electrolyte also limits the choice of stable plate materials [A 146].

The more widely employed solutes contain SCN^- as an anion. An important side reaction, then, is the electrochemical decomposition of thiocyanate. The reactions below [determined *versus* $Pb/Pb(NO_3)_2$] were recently determined for 21 molal NaSCN [A 132]:

Cathode	$SCN^- + 2e^- \rightarrow S^{-2} + CN^-$	$E = -1.75$ V
Anode	$2SCN^- \rightarrow (SCN)_2 + 2e^-$	$E = +1.15$ V

Thus, the maximum available cell voltage with this solute is 2.90 V. These reactions were determined on a platinum surface, which would probably tend to minimize any overvoltage effects. The potential of $+1.15$ V is very close to the decomposition potential for liquid ammonia, so that a mixed potential may actually have been obtained [A 253]. From a preliminary study of properties of some liquid NH_3 electrolytes, it is concluded that the stability range of NH_3 is more probably -1.00 to $+1.35$ V *versus* $Pb/PbCl_2$. Further details of liquid NH_3 chemistry are given elsewhere [A 108–110].

Negatives

Because of a lower degree of dissociation, liquid ammonia is less easily reduced by active metals than is water. It is thus possible to use magnesium anodes with relatively few corrosion problems, at least in neutral or basic electrolytes.

The more active metals, such as lithium, sodium, potassium, and calcium, are soluble to a significant extent in liquid ammonia [A 99,105,106; P 341,390]. At temperatures below about $-40°C$, the alkali-metal–ammonia solutions separate into two phases [A 105,107]. The dilute phase is blue in color, and its spectra are nearly independent of the nature of the metal. It contains metal ions and solvated electrons and is a very strong reducing agent. The more concentrated phase is brown or bronze in color with a metallic luster and is highly conductive, like a metal. Test cells have been built employing lithium as the negative [A 255]. In comparison tests, it was found that the lithium cell potential was 0.4 V higher at 100 mA/cm² than that of a magnesium cell. Furthermore, the magnesium cell had a 72% decay in 5 min, compared with 35% for the lithium cell. Under similar conditions, it was also found that the lithium cell activated approximately twice as fast as the magnesium cell.

Positives

The cathodes which have been employed in primary cells using liquid ammonia electrolytes are similar to those used in aqueous electrolytes, *e.g.*, mercurous sulfate and silver chloride [A 101]; however, as shown previously, these materials are not high-energy compounds. Mercurous sulfate also forms a stable adduct with ammonia, and the heat of reaction upon the addition of electrolyte has compli- cated battery operation [A 135]. Chemical analysis of the spent electrolyte and gases evolved during life tests implies that mercu- rous sulfate also decomposes the electrolyte, evolving nitrogen [A 254].

Sulfur, a high-capacity oxidant, is fairly soluble in liquid ammonia; it forms species such as S_2N_2 and S_4N_4 as well as ammonium sulfide. This solubility has caused problems with using sulfur-based cathodes in batteries, since migration of soluble sulfur species to the anode results in direct oxidation of the magnesium anode material to MgS [A 97].

It has been observed [A 355] that the addition of sulfur to the $HgSO_4$ cathode significantly improved performance; the optimum $S/HgSO_4$ ratio was 2. The exact mechanism for the improvement is not known; the following reaction mechanism was suggested:

$$HgSO_4 + 2KSCN + 2S \rightarrow Hg + K_2SO_4 + S_2(SCN)_2$$

where $S_2(SCN)_2$ is the electroactive material.

Meta-dinitrobenzene has been studied extensively. This compound is reduced to m-bis-hydroxylaminobenzene in an 8-electron reaction, similar to its reaction in aqueous solution. Ortho- and p-DNB may provide higher energy densities because they undergo a 12-electron reduction to the diamino benzene [A 104]. As in aqueous systems, these reductions are most efficient in acid electrolytes, which is reasonable in that protons are consumed in the cathode discharge reaction. Ortho- and p-DNB have large voltage steps in reduction; hence, voltage regulation is poorer at low rates. However, at high rates, m- and p-DNB give similar voltage regulation [A 104].

All the depolarizers listed above are nonconductors; therefore, a conducting material, generally carbon, is included in the cathode mix. The mix ratio affects battery performance, internal battery temperature, and pressure [A 136]. The mix containing the largest amount of carbon discharges at a higher voltage and a lower internal pressure and tem- perature. (A discussion of working NH_3 electrolyte batteries is given in Chapter 6, pp. 230-233.)

LIQUID SULFUR DIOXIDE

Liquid SO_2 is an aprotic, inorganic solvent which is also relatively stable in the presence of alkali metals. It forms a conducting electrolyte with some solutes; *e.g.*, a concentrated solution of KI has a conductivity of 8.95×10^{-2} $(\Omega\text{-cm})^{-1}$ at $0°C$. Unlike liquid NH_3, solutes do not appreciably lower the vapor pressure of SO_2. As a consequence, it is necessary to work under pressure or below the normal boiling point of the mixture $(0–10°C)$.

A Na/IBr_3 cell has been built; an open circuit voltage of 3.0–3.50 V was observed [A 141]. There is also some indication that high current densities are possible [A 142]. A thin battery of the type $K/KCl \cdot SbCl_5$ (0.25M), SO_2/IBr_3 + carbon had an open circuit voltage greater than 3.70 V and an initial voltage of 3.15 V, decreasing to 2.40 V in 10 min at a current density of $20\ mA/cm^2$. Polarizations, such as activation (charge transfer), concentration, and *IR* losses in the electrolyte, were shown to be of minor consequence. It was determined that the surface film on the anode and cathode was the major cause of polarization [A 309].

ANHYDROUS HYDROGEN FLUORIDE

Hydrogen fluoride is a liquid in the temperature range from -83.1 to $19.54°C$. The specific conductivity is 1.6×10^{-6} $(\Omega\text{-cm})^{-1}$ at $0°C$ [A 137]; the exact value is very much a function of the water content. The conductivity can be increased substantially by solutes of NaF, KF, or NH_4F; liquid HF is a fairly good solvent, being similar to H_2O and NH_3.

Because of the availability of labile hydrogen, this solvent cannot be used with alkali metal anodes. Concentrated HF solutions and metals such as titanium may also be incompatible because of corrosion reactions.

The fluorine electrode itself has been reported to be irreversible [A 138]; the emf of the cell H_2/HF, KF/F_2 was reported to be 2.768 V on platinized platinum electrodes at $0°C$. Large overvoltages have been observed when producing fluorine from KF–HF melts by electrolysis [A 139].

The $Ni–NiF_2$ couple was also observed to be irreversible [A 140]. In addition, the electrode potentials of the system studied, although theoretically fluoride-independent, were noticeably affected by the ionic strength of the electrolyte solutions. The cell Cu, CuF_2/TlF, HF/$TlF_3(s)$ Pt was found to be reversible and reproducible, having a standard emf of 0.9282 V at 0°C [A 138]. Apparently, the $AgF(s)$/Pt electrode is also reversible [A 138].

FUSED SALT ELECTROLYTES

The temperature range of application of fused salt electrolytes is compared to that of aqueous and other nonaqueous electrolytes in Fig. 3-3 [A 143].

Batteries employing this type of electrolyte are generally referred to as "thermal" batteries because of their heat-activation characteristic. The cells are stored at ambient temperature, with the electrolyte a solid. This provides for a low self-discharge rate and a long storage life. When fused, the cells are capable of high discharge rates for short times. It has been in this area of high discharge rates ($\geqslant 1$ A/in.²) that the thermal battery has found most application. Figure 6–2 gives a comparison of the high-rate performance of a molten salt electrolyte battery with other systems [A 143].

Electrolyte Properties

Ionic species are not hydrated in fused salt solutions, so that the thermodynamic uncertainties and kinetic effects produced by

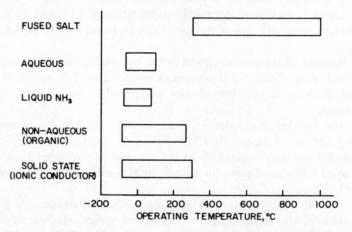

Fig. 3-3. Typical temperature ranges for various electrolytes [A 143].

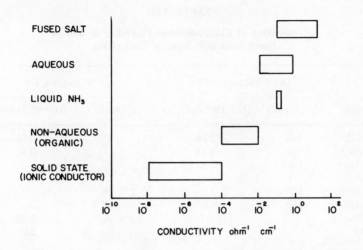

Fig. 3-4. Typical conductivity ranges for various electrolytes [A 143].

solvation are essentially absent [A 152]. The conductivities of molten salts are from 10 to 100 times higher than those of aqueous systems, so that molten salt cells should have low voltage losses due to the *IR* drop. (Conversely, given a highly conductive electrolyte, it should be possible to discharge at higher rates before attaining an *IR* loss comparable to that achieved in aqueous systems.) The use of high currents requires that considerable attention be paid to elimination of ohmic resistance in other parts of the battery, *e.g.*, contact resistance within the leads. Figure 3-4 gives a comparison of the conductivity of fused salt systems with that of other electrolytes [A 143].

Diffusion constants for ions in the LiCl–KCl eutectic (450°C) are also compared (Table 3-III) to similar ions in aqueous systems. It is to be noted that the differences are only a factor of 2 or 3. Thus, diffusion rates in molten salts are not greatly different from those in aqueous electrolytes. This fact, considered together with the high reaction rates, indicates that the rate-limiting step in the discharge reactions should be mass transport. Table 3-III also gives a listing of exchange currents for selected couples in a KCl–LiCl eutectic at 450°C [A 152]. For comparison purposes, the exchange currents of some couples in aqueous solutions (25°C) are included [A 153].

At corresponding temperatures relative to the melting point, simple ionic salts do not possess physical properties radically different from

TABLE 3-III

Comparison of Electrochemical Properties of Ions in Fused Salts and Aqueous Electrolytes

| Couple | KCl–LiCl (450°C) | | H$_2$O (25°C) | |
	i_0* (A/cm^2)	D ($\times 10^{-5}$ cm^2/sec)	i_0† (A/cm^2)	D ($\times 10^{-5}$ cm^2/sec)
Cd/Cd^{+2}	210	2.08		0.72
Pb/Pb^{+2}	30	2.18		0.98
Ag/Ag$^+$	190	2.6		
Bi/Bi^{+3}	8	0.6		
Zn/Zn^{+2}	150		2×10^{-5}	
Ni/Ni^{+2}	110		2×10^{-9}	
Cu/Cu^{+2}		3.5	2×10^{-5}	0.72
Tl/Tl$^+$		3.88		2.0

* For 1 M solution.
† For 1 M solution of the sulfate.

TABLE 3-IV

Properties of Water and Molten Sodium Chloride at θ = 1.06*

Substance	NaCl (850°C)	H$_2$O (16°C)
Surface tension (dynes/cm)	110.8	73.3
Viscosity (cP)	1.20	1.11
Density (g/cc)	1.5295	0.9989
Vapor pressure (mm Hg)	0.89	13.63

* Taken from [A 154].

other liquids. For example, Table 3-IV lists the properties of molten NaCl (m.p., 801°C) and water, both at $\theta = T/T_{m.p.} = 1.06$.

Although density and surface tension of fused sodium chloride are larger by a factor of 1.5, the viscosities are nearly identical. Low vapor pressure and high surface tension of fused sodium chloride reflect strong cohesive forces within the liquid. Table 3-V lists the composition and melting points of mixtures frequently used.

Molten alkali chloride mixtures have been used as battery electrolytes. Chloride is not readily oxidizable and alkali metals are not

TABLE 3-V

Composition and Melting Points of Some Fused Salt Mixtures
(Component C by Difference)

Type of Melt (A–B–C)	Mol.% A	Mol.% B	Melting point (°C)
LiCl–KCl eutectic	59	41	352
NaCl–KCl minimum	50.1	49.9	658
Li_2SO_4–Na_2SO_4–K_2SO_4 eutectic	78	8.5	512
Li_2SO_4–K_2SO_4 eutectic	80	20	535
Li_2CO_3–Na_2CO_3–K_2CO_3 eutectic	43.5	25.0	397
Equimolar KNO_3–$NaNO_3$	50	50	220
$LiNO_3$–$NaNO_3$–KNO_3 eutectic	30	17	120
$LiNO_3$–NH_4NO_3–NH_4Cl melt	25.76	66.65	86.2
KCl–$AlCl_3$ eutectic	33	67	128

easily reduced, so that a polarization span of about 3.6 V exists between the oxidation of chloride and the reduction of alkali. The molten alkali halide mixtures are all rather high-melting, except for those containing lithium chloride; complete dehydration of these melts is difficult. Since water and chloride ion react to produce HCl and hydroxide ions in the fused melts, all water must be removed prior to use.

Molten sulfates have received relatively little attention in the past. In the Li_2SO_4–K_2SO_4 eutectic, at 600°C, sulfate ion is oxidized at +0.7 V [versus a Ag–Ag (I) reference electrode] to oxygen and sulfur trioxide. Reduction of sulfate to sulfite, sulfide, and sulfur occurs at −1.6 V. The available voltage span in fused sulfates is 2.3 V, significantly smaller than that in fused chlorides.

Molten carbonates are chiefly of interest in high-temperature fuel cells. Their extremely corrosive nature makes experimental study difficult.

Molten nitrates have received much attention due to their low melting points and ease of handling. Although approximately 5 V is available between the complete oxidation of nitrate and reduction of alkali in fused sodium–potassium nitrate, a reduction process involving nitrate ion occurs at approximately −1.5 V versus a Ag–Ag (I) reference [A 156,185]. Studies carried out at 250°C demonstrate that the peak phenomenon observed at approximately −1.65 V in cathodic polarization curves using various solid microelectrodes and dropping mercury may be identified with the beginning of the irreversible reduction of NO_3^- to NO_2^-, limited by the precipitation of Na_2O at the electrode surfaces. It appears that the after-peak current limiting process is the rate of dissolution of the Na_2O film, which determines the rate of reduction necessary to sustain a steady state. The anodic and cathodic limits of potential are observed to represent the evolution of NO_2 gas (-1.2 V) and reduction of alkali metal (-2.8 V); subsequent reaction of deposited alkali metal with NO_3^- produces a colorless, odorless gas and oxide ion in amounts indicating nitrogen as the reduction product [A 150,156].

The alkali halide–aluminum halide eutectics are also of interest because of their low melting points. The eutectics consist of AlX_4^-–M^+, AlX_3, and Al_2X_6 species. The ease with which aluminum halides react with water makes it difficult to prepare these melts without some contamination.

A significant number of salt mixtures are immiscible; e.g., 144 insoluble binary mixtures were known in 1960, including most of the alkali nitrate–silver halide combinations [A 154].

One of the principal problems with fused salt systems is the necessity of maintaining the high operating temperature. Other factors being equal, a smaller differential between storage ambient and cell operating temperature would unquestionably improve the life of fused salt cells. The efficiency of heat transfer decreases with a decreasing temperature differential. Several surveys [A 257–259] have been carried out with the goal of a low-melting electrolyte of high ionic conductivity and good thermal stability. These studies included various fluorides, fluorophosphates, fluoroborates, and acetamide. Except for the latter, these electrolytes were unsuitable for use as low-temperature fused salts. Magnesium reacts with molten acetamide, generating hydrogen and limiting the utility of this solvent.

Acetamide melts at 80°C and boils at 222°C. Molten acetamide possesses a viscosity of 1.32 cP at 105°C and 1.06 cP at 120°C [A 261]; the specific conductivity is approximately 4×10^{-6} $(\Omega\text{-cm})^{-1}$ at 94°C [A 262]; the dielectric constant is 60.6 at 94°C. The solubilities in inorganic compounds are analogous to those in water. Exceptions are $KClO_4$ and the halides of mercury, which are more soluble in molten acetamide than in water [A 260]. Zinc, tin, lead, and cobalt could be electrodeposited. Attempts to deposit the more reactive metals, magnesium and aluminum, were unsuccessful [A 263]. Hydrogen evolution was noted from a magnesium negative in acetamide. As with water, a negative difference effect was observed. A similar result was obtained with $Al/AlCl_3 \cdot$ acetamide. The cell Mg/acetamide, 21% $LiCl/V_2O_5-B_2O_3$, Ni had at $250°C$ an open circuit potential of 2.81 V and sustained a current density of 80 mA/cm^2 at a voltage of 1.32 V [P 420].

A stable cell was constructed from [A 264]:

$$Zn/KCl, acetamide/AgCl, Ag \qquad E = 1.1 \text{ V}$$

No abnormalities were observed on discharging or charging the cell. The electrolyte behaves substantially as a high-temperature analog of water, although there was some difficulty at high rates. The influence of design could not be evaluated, since exchange currents were not measured.

Another class of material is the organic salts, *e.g.*, sulfamates, formates, acetates, and substituted pyridinium salts [A 264]. Ethyl-pyridinium bromide has a conductivity of 7×10^{-2} $(\Omega\text{-cm})^{-1}$ at $170°C$. It was found that silver halides are soluble, while chromates decomposed the salt. PbO_2 is virtually insoluble and inert in the melt for 1 hr at a temperature of $200°C$. The following cell was formed and then operated at $200°C$: $Mg/methylpyridinium$ $bromide/CaCrO_4$, Ni. Appreciable loads could not be drawn as a result of insoluble reaction product formed at the negative.

Electrode Potentials

As mentioned in Chapter 1, many electrode reactions are reversible in fused salt reactions. Hence, there is generally a linear dependence of overpotential on current, and the back reaction must be considered in interpreting half-cell potentials.

Reference Electrodes. An extensive discussion on fused salt reference electrodes is available [A 158]. Silver–silver-ion couples have been widely employed in fused salt studies. A silver wire in contact with a solution of silver ions of known concentration in the solvent and separated from the bulk melt by a conductive barrier has been generally useful. In some cases, a silver wire–pure silver salt system, such as Ag/AgCl, has been used, but the liquid junction potential between reference electrode and salt solution should be larger in this case [A 158]. A platinum–platinous chloride couple has also been popular in fused chlorides at moderate temperatures. $PtCl_2$ decomposes at high temperatures, and this places a limitation on the utility of this system. Sodium–tin alloys and sodium–mercury amalgams in conductive glass containing sodium ions have been popular with Russian workers.

Standard Potentials. Table 3-VI lists half-cell potentials for the LiCl–KCl eutectic at 450°C referred to the Pt–Pt (II) reference electrode [A 167]. No claim is made for the reversibility of all systems indicated, particularly the oxides. The more important battery couples are discussed in a later section.

Half-cell potentials in a KCl–NaCl electrolyte at various temperatures are given in Table 3-VII. Listings of standard half-cell potentials in other salt systems are less extensive (Tables 3-VIII and 3-IX). To express concentration, three scales can conveniently be used. These are molarity, molality, and mole fraction. If the assumptions are made that the quantity of the solute is negligible compared to that of the solvent and that the density of the solution is the same as that of the solvent, it can be shown that

$$M = \frac{W_s d}{M_s N M_0} \times 10^3$$

$$m = \frac{W_s}{M_s N M_0} \times 10^3$$

$$X = \frac{W_s}{M_s N}$$

where M is the molarity; m is the molality; X is the mole fraction; W_s is the weight of the solute (g); M_s is the formula weight of the solute; M_0 is the average formula weight of the solvent; N is the number of moles of chloride in solution; and d is the density of the solvent (g/ml). At 450°C, the density of the eutectic (KCl—LiCl) melt is 1.648 g/ml [A 168] and the conversion factors are

$$X = 0.0337 \, M$$

and

$$m = 0.607 \, M$$

The advantage of molality and mole fraction over molarity is that the first two do not involve the density of the melt and are independent of temperature.

TABLE 3-VI

Standard Potential Series for LiCl–KCl at 450°C *

Electrode system	Half-cell potentials E^0 (V)		
	Molarity	Molality	Mole fraction
Ca(II)–Ca(0)	−3.17[†]		
Li(I)–Li(0)	−3.304	−3.320	−3.410
Mg(II)–Mg(0)	−3.580	−2.580	−2.580
Mn(II)–Mn(0)	−1.849	−1.849	−1.849
Al(III)–Al(0)	−1.762	−1.767	−1.797
Zn(II)–Zn(0)	−1.566	−1.566	−1.566
V(II)–V(0)	−1.533	−1.533	−1.533
Ti(I)–Ti(0)	−1.476	−1.460	−1.370
Cr(II)–Cr(0)	−1.425	−1.425	−1.425
Cd(II)–Cd(0)	−1.316	−1.316	−1.316
NiO–Ni	−1.23[‡]		
Cu_2O–Cu	−1.207[‡]		
Fe(II)–Fe(0)	−1.171[§]	−1.171	−1.171
Pb(III)–Pb(0)	−1.101	−1.101	− 1.101
Sn(II)–Sn(0)	−1.082	−1.082	−1.082
Co(II)–Co(0)	−0.991	−0.991	−0.991
Cu(I)–Cu(0)	−0.957	−0.941	−0.851
α-Fe_2O_3	−0.88[†]		
Ga(III)–Ga(0)	−0.84	−0.84	−0.88
In(III)–In(0)	−0.800	−0.805	−0.835
Ni(II)–Ni(0)	−0.795	−0.795	−0.795
V(III)–V(II)	−0.748**	−0.764	−0.854
Ag(I)–Ag(0)	−0.743	−0.727	−0.637
Sb(III)–Sb(0)	−0.635	−0.640	−0.670
Bi(III)–Bi(0)	−0.553	−0.558	−0.588
Cr(III)–Cr(II)	−0.525	−0.541	−0.631
Hg(II)–Hg(0)	−0.5	−0.5	−0.5
PdO–Pd	−0.514[‡]		
V_2O_5	−0.4[‡]		
PtO–Pt	−0.338[‡]		
Pd(II)–Pd(0)	−0.214	−0.214	−0.214
I_2–I^-	−0.207	−0.254	−0.524
Pt(II)–Pt(0)	0.000	0.000	0.000
Cu(II)–Cu(I)	+0.061	+0.045	+0.045
Br_2–Br^-	+0.177[§]	+0.130	−0.140
Au(I)–Au(0)	+0.205	+0.221	+0.311
Cl_2–Cl^-	+0.322	+0.306	+0.216

* All entries taken from [A 167] except as otherwise noted.
† Taken from [A 162].
‡ Taken from [A 161].
§ Taken from [A 169].
** Taken from [A 166].

TABLE 3-VII

1:1 KCl–NaCl at High Temperatures*
$E^0(M/M^{+n}) - E^0$ (Ag/Ag$^+$)

Electrode	Half-cell potentials (V)		
	700°C	800°C	900°C
Mn/MnCl$_2$	−1.206	−1.190	−1.172
Zn/ZnCl$_2$	−0.860	−0.835	−0.810
Cr/CrCl$_2$	−0.758	−0.740	−0.728
Tl/TlCl	−0.665	—	—
Cd/CdCl$_2$	−0.620	−0.580	—
Fe/FeCl$_2$	−0.520	−0.510	−0.498
Cr/CrCl$_3$	−0.425	−0.385	−0.345
Pb/PbCl$_2$	−0.390	−0.376	−0.355
Sn/SnCl$_2$	−0.370	−0.354	−0.340
Co/CoCl$_2$	−0.324	−0.300	−0.275
Cu/CuCl	−0.260	−0.256	−0.260
Ni/NiCl$_2$	−0.140	—	—
Ag/AgCl	0.0	0.0	0.0
Cu/CuCl$_2$	+0.170	+0.180	+0.192
Cl$_2$/Cl$^-$	+0.845	+0.820	+0.795

* Taken from [A 170].

TABLE 3-VIII

Standard Electrode Potentials in Molten
MgCl$_2$–NaCl–KCl at 475°C*

Electrode system	Standard electrode potentials (V)	
	Molality	Mole fraction
Cr(II)/Cr(0)	−1.396	−1.396
Fe(II)/Fe(0)	−1.183	−1.183
Cr(III)/Cr(0)	−1.131	−1.159
Cu(I)/Cu(0)	−0.947	−0.863
Fe(III)/Fe(0)	−0.852	−0.880
Cr(III)/Cr(II)	−0.602	−0.685
Cu(II)/Cu(0)	−0.519	−0.519
Fe(III)/Fe(II)	−0.190	−0.274
Cu(II)/Cu(I)	−0.091	−0.175
Pt(II)/Pt(0)	0.000	0.000

* Taken from [A 187].

TABLE 3-IX

Standard Potentials in Molten NaF–KF at 850°C*

Couple	Potential (V)
Al(III)/Al	−1.5
Mn(II)/Mn	−1.04
Cr(III)/Cr	−0.70
Fe(III)/Fe	−0.12
Co(II)/Co	−0.07
Ni(II)/Ni	0.00 (standard)
Cu(I)/Cu	+0.48
Ag(I)/Ag	+0.64

* Taken from [A 188].

"The standard state for a metal ion is unit concentration except for Li+, for which the prevailing activity in the eutectic mixture was adopted as the standard state. For pure metals, the standard state is defined as its physical state at 450°C under one atmosphere pressure and taken as unity. The standard potential of an electrode system is its potential at unit ratio of oxidant to reductant activity. Thus, for the three concentration scales, there are three standard potentials for each electrode system. These are designated as E_M^0, E_m^0, and E_X^0 for the molarity, molality, and mole fraction scales, respectively" [A 167].

Cell Materials—General

A limited table of solubilities is given in Table 3-X [A 157]. As a rule-of-thumb, all materials are soluble to some extent in fused salts.

Apparently, lithium does show the least tendency of all alkali metal systems to mix with its halides in the liquid phase. The solubility of lithium in LiCl at 640°C was 0.5–0.2 mol.% and that in LiI, 1.2 mol.% at 550°C.

A number of surveys have been made of possible plate materials, e.g., under Contract Nord 18240 and under Contract AF 33(616)-7505. It would serve little purpose to report here the listings of theoretical performance for proposed battery couples. Instead, the discussion will be concerned with general principles and will describe the performance of some of the more successful couples.

TABLE 3-X

Solubility of Metals in Fused Halides*

Metal	Salt	Temperature (°C)	Solubility of the metal (mol.%)	Metal	Salt	Temperature (°C)	Solubility of the metal (mol.%)
Na	NaF	1000	7.6	Ba	$BaCl_2$	1050	30.60
Na	NaCl	811	2.8	Ba	$BaBr_2$	1050	36.70
Na	NaCl	1000	33.0	Ba	BaI_2	1050	39.40
Na	NaBr	720	0.15	Zn	$ZnCl_2$	500	8.9×10^{-5}
Na	NaBr	943	14.7	Cd	$CdCl_2$	600	15.20
Na	NaI	722	4.1	Cd	$CdCl_2$	690	16.40
Na	NaI	954	20.8	Cd	$CdBr_2$	600	13.90
K	KF	948	23.3	Cd	CdI_2	600	6.07
K	KCl	800	7.6	Hg	$HgCl_2$	350	50.0
Cs	CsF	692	6.0	Hg	HgI_2	350	33.6
Cs	$CsCl_2$	626	9.0	Ag	AgCl	700	0.0600
Cs	CsI	597	28.0	Al	AlI_3	423	0.3000
Mg	$MgCl_2$	800	1.08	Ga	$GaCl_2$	180	1.9200
Mg	$MgCl_2$	1050	1.57	Tl	TlCl	550	0.0090
Ca	$CaCl_2$	1000	5.40	Sn	$SnCl_2$	500	0.0032
Ca	CaI_2	1000	9.66	Sn	$SnBr_2$	500	0.0680
Sr	SrF_2	1000	19.90	Pb	$PbCl_2$	600	0.0200
Sr	$SrCl_2$	1000	24.60	Pb	$PbCl_2$	800	0.1230
Sr	$SrBr_2$	1000	36.10	Pb	$PbCl_2$	700	0.0550
Sr	SrI_2	1000	39.30	Bi	$BiCl_3$	450	47.50
Ba	BaF_2	1050	21.90	Ce	$CeCl_3$	850	33.00

* Taken from [A 157].

The electrolyte most often used has been the LiCl–KCl eutectic at 450–500°C; some attention has been given to the K_2SO_4–Li_2SO_4 eutectic and the $CuPO_3$–$LiPO_3$ system at 600–900°C. These electrolytes have been used principally in conjunction with SO_4^{-2} and PO_3^- depolarizers.

A number of suitable anode materials are available, *e.g.*, calcium and magnesium, and thermal batteries have been constructed with these elements. As with most other battery systems, the principal problem area is that of a stable and active positive material. A complication arises from the significant solubility of most salts and oxides in fused salt media. Thus, active material is lost to the electrolyte and plates out on the negative. If the plated metal is insoluble, the negative can be effectively coated with an inactive material. This is often the case when the active material is an anion oxide, *e.g.*, CrO_4^{-2}; an insoluble mixed oxide often forms, deactivating the negative. Most often this effect is ignored, with the idea that the loss in performance will be small for the short period of time that the battery must function, *i.e.*, 5–15 min. A listing of solubilities of positives is given in Table 3-XI.

As with aqueous systems, the use of membranes has been suggested to relieve this problem of chemical short-circuiting. One attempt at this [A 189] was the use of ceramic permeable only to sodium ion. It was not possible, however, to develop a structure with sufficient permeability and low resistivity suitable for high drain rates. A second attempt [A 190] involved the use of a thin, synthetic zeolite membrane. The pure, hot-pressed, heat-treated (700°C) NaX-type zeolite had a conductivity

TABLE 3-XI

Solubility of Cathodic Materials in Halide and Sulfate Electrolytes*

Cathodic material	Solubility(%)	
	KCl–LiCl at 600°C	K_2SO_4–Li_2SO_4 at 800°C
CrF_3	1.96	0.21
CoF_3	Reacts	4.33
$CuCl_2$	11.4	2.1
CuO	0.36	0.11
Fe_2O_3	0.04	0.06
K_3FeF_6	15.0	4.89
MnF_3	4.32 With reaction	3.57
MoS_2	0.42	5.78
Na_2CrO_4	13.5	6.3
Sb_2O_3	0.50	0.10
V_2O_3	17.0	2.58
WO_3	3.24	9.05
	KCl–LiCl at 450°C[†]	
Cu_2O	0.42	
PtO	0.09	
PdO	0.11	
NiO	~0.002	

* Taken from [A 171].
† Taken from [A 167].

of 2×10^{-3} $(\Omega\text{-cm})^{-1}$ at 510°C. Further improvements are necessary for high drain rates. For example, the ohmic loss within a 1-mm-thick section of this electrolyte would be 0.5 V at 100 mA/cm².

Oxide-deficient materials, *e.g.*, $(ZrO_2)_{0.85}(CaO)_{0.15}$, have been used as solid electrolytes in high-temperature fuel cells [A 190]. The CaO system forms cubic crystals of the fluorite type, with all cation lattice sites occupied, but with one anion site vacant from each CaO molecule included. Transfer of oxide ions from one vacant anion site to another accounts for the conductivity of these materials. However, the temperature must be rather high (about 1000°C) before conductivity becomes high enough for practical use.

Since thermal batteries are intended for high-rate discharge, it is desirable that the electrochemical half-cell reaction products be soluble. This is generally not a problem with the negative material, since the chlorides of magnesium and calcium are quite soluble in the LiCl–KCl eutectic. However, at high rates, $CaCl_2$ will precipitate on the electrode surface, thereby reducing the conducting path at this interface and increasing ohmic polarization [A 264]. In the case of simple oxide positives, *e.g.*, Cu_2O, this also need not be a problem. The discharge deposits O^{-2} into the electrolyte and metal on the electrode. The metal is of a higher density than the original oxide, so that the porous plate structure should remain. Furthermore, the metal should improve electrode conductivity. Fortunately, the oxides of magnesium and calcium are somewhat soluble, so that little precipitation occurs within the electrolyte. The problem becomes more complicated when oxide anions are used. The product is often a lower oxide which is insoluble and can coat the electrode.

Theoretically, the optimum battery couple for a LiCl electrolyte is Li/Cl_2. (KCl is often omitted from the electrolyte to decrease the solubility of molten lithium.) In principle, this system should lend itself to secondary battery applications for the same reasons, and concentration polarization on charging should be negligible. However, a number of serious engineering problems are involved in developing such a device, arising from the fact that lithium metal is molten at the temperature of operation, and LiCl can be quite corrosive.

A list of *general problems in constructing thermal batteries* is as follows:

1. Residual gases retained in the battery will expand, causing a separation of materials.

2. Moisture retained by the materials forms steam which has the same effect as the expanding air and, in addition, oxidizes the anode prematurely.
3. The liquid phases of the anode and cathode migrate across the electrolyte causing internal shorts.
4. Carbon binder in the cathode migrates in the same manner as the liquid salts and metal eutectics, causing additional shorting.

In an idealized configuration, where these problems were eliminated, the service life of the following battery was measured in terms of days at 800–1000°F [A 363]:

$$Mg/LiCl–KCl/Cu_2O + CuO$$

Faradaic efficiencies of 73–93% were achieved, as well as an energy density of 28 W-hr/lb (excluding packaging).

Negatives

Calcium. This material has been used as a negative electrode in a number of thermal batteries [A 159,160]. The potential, with respect to a KCl–AgCl glass reference electrode, was reported as −3.0 V [A 159]. The equivalent weight of calcium is 20.

When operated in an electrolyte of $BaCl_2$–NaCl–$CaCl_2$ at 500°C, it is observed that the electrode is slow to activate. This delay in equilibrium with electrolyte was attributed to the formation of a layer of calcium oxide and nitride on the surface during cell assembly. Apparently, the dissolution of CaO is a slow process in this system. In the similar LiCl–KCl eutectic, it is known [A 161] that CaO is fairly soluble and can actually be used as a source of oxide ion. Furthermore, it is possible, in a lithium electrolyte, for the following reaction to take place:

$$Ca^0 + 2Li^+ \rightarrow Ca^{+2} + 2Li^0$$

The lithium metal so formed will develop a liquid alloy with calcium at temperatures above 231°C. It was postulated that this fluid surface, continuously renewed by convection flow, would not be subject to the marked polarization developed by the wholly solid surface. Indeed, fast activation times and low polarization at high rates are observed [A 159]. However, a second problem is introduced—that of structural stability. It is possible for the fluid negative material to migrate through the electrolyte compartment and discharge at the positive plate. Apparently, this effect can become quite pronounced at higher temperatures. There is contained, in these reactions, a mechanism for thermal runaway, since both reactions—the alloy formation and the self-discharge reaction—are exothermic. This reaction can limit cell life to 5–10 min.

Some dimensional stability was achieved by including in the electrolyte a species which is an oxidizing agent and, at the same time, a source of oxide ion. Several of these, including permanganate, m-vanadate, and nitrate ions, were tried, but chromate ion (6.65 mol.%) was the most satisfactory in terms of giving an electrode with a very stable anodic potential and a long discharge life. A somewhat insoluble and protective coating containing calcium and Cr_2O_3 developed on the discharged calcium electrode. Apparently, this film was sufficient to suppress, but not completely inhibit, the lithium alloy reaction. A cell employing this system was discharged for 11 min at 1.50 A/cm² and at an anode potential of -2.59 V (versus the Ag/AgCl electrode). However, it can be difficult to maintain the proper balance of chromate to inhibit alloy formation and not inhibit electrochemical discharge.

Magnesium. This metal and calcium are the most widely used negative electrodes in thermal battery systems [A 162,163]. The equivalent weight of magnesium is 12; the half-cell potential is approximately 0.3 V less negative than a calcium electrode operating in a LiCl–KCl electrolyte [A 164]. Unlike the case of the calcium electrode discussed previously, a lithium-ion electrolyte is not needed for optimum discharge of magnesium.

The dissolution of magnesium on open circuit in LiCl–KCl at 500°C has been determined as $1.0 \pm 0.2 \times 10^{-6}$ g (8×10^{-8} equivalents) per cm² of exposed surface per minute. At reasonable discharge rates, this side reaction is not considered important; electrode efficiency is generally greater than 91% [A 162].

Essentially, all of the oxidized magnesium is found in the electrolyte and in a water-soluble form. Presumably, the half-cell reaction is as follows:

$$Mg + 2Cl^- \rightarrow Mg^{+2} + 2Cl^- + 2e^-$$

If the electrolyte itself contains oxide ion or hydroxide, either as impurity or reaction product, it is possible to form significant amounts of MgO. This has been observed when unpurified LiCl or KCl is used.

The water present in these salts supplies the oxygen necessary for the formation of insoluble MgO [A 165].

Aluminum. The equivalent weight of this material is 9. A high-temperature cell has been described [A 280] which has an aluminum negative and an O_2/Cu positive. The electrolyte consisted of 40.5% AlF_3, 57.85% NaF, and 2.65% Al_2O_3.

Lithium. The equilibrium potential for the Li(I)–Li(0) couple in the LiCl–KCl eutectic is listed in Table 3-VI. The determination of this potential presented special problems, since molten lithium in the presence of the chloride melt is extremely reactive and attacks glass very rapidly.

This metal possesses two properties which made it desirable for use as a negative electrode in a thermal battery—a high potential and a low equivalent weight (1750 A-hr/lb). Demonstration cells employing lithium negatives have been constructed [A 172] and operated on charge and discharge cycles. A performance curve for the Li/AgCl couple is shown in Fig. 3-5; the lithium electrode is shown in Fig. 3-6.

The lithium electrode has also been used in the $Li/LiH/H_2$ thermally

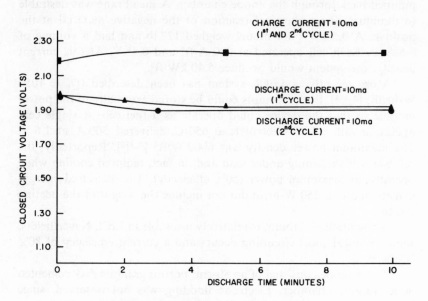

Fig. 3-5. Charge and discharge data for Li–AgCl/Ag couple.

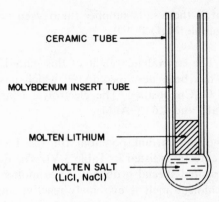

Fig. 3-6. Molybdenum insert tube arrangement for Li–AgCl/Ag couple.

regenerative fuel cell [A 173]. In the cell reaction, lithium and hydrogen form LiH. A fused salt of LiCl–LiF eutectic (m. p., 932°F) was used as electrolyte. The LiH product dissolved in the eutectic; this solution was removed from the cell and pumped to a regenerator (1650°F). At this temperature, the LiH decomposed to lithium and hydrogen, which distilled from the melt for recycling. Lithium and electrolyte were pumped back through the anode chamber. A membrane was desirable to minimize direct chemical reaction of the negative material at the positive. A 0.5-kW pilot model weighed 127 lb and had a volume of 0.688 ft³. Each cell operated at 200 A/ft² and 0.45 V. At this current density, the system would produce 5.40 kW/ft³.

More recently, a Li/Cl_2 system has been described ($Q_0 = 1050$ W-hr/lb) [A 174]. Both reactants could be stored externally to the power module so that the device could operate as a fuel cell. A single cell, operating with a LiCl electrolyte at 650°C, delivered 300 A at 1.6 V. The maximum power density was 6000 W/ft² [A 175]. Reportedly, the cell was self-sustaining under load and, in fact, required cooling when operated at maximum power (50% efficiency). The measured energy-density figure of 250 W-hr/lb did not include the weight of the heating system.

As mentioned, lithium is relatively insoluble in LiCl. Nevertheless, some chemical short-circuiting occurs and a current efficiency of 80% is observed.

The cathode consisted of an 80-cm² porous graphite disk cemented in a graphite chamber. Electrode flooding was not observed, since LiCl does not wet graphite. The lithium anode consisted of a stainless-steel fiber metal disk in an iron chamber. A beryllium oxide ring served as an electrical insulator. This cell was designed with side feeding, which allows stacking of cells into a battery.

The success of this type of system is dependent upon solving severe materials problems similar to those which complicated development of the Li/H_2 cell. Of course, lightweight electrode structures and separator membranes are also necessary.

Lithium Alloys. A commercially available lithium–aluminum–magnesium alloy has also been used in thermal cells [A 191]. This alloy contains 13–15% Li and 1.0–1.5% Al. It reportedly gives cell voltages both on open circuit and on drain 0.55 V higher than pure magnesium [$V_{oc} = -3.10$ V *versus* Pt/Pt (II)]. The particular advantage of this alloy is its high electrochemical activity without the formation of a liquid anode. It is also claimed that the alloy does not react with water even at 210°C and is unaffected by contact with the molten LiCl–KCl eutectic.

Lithium Hydride. The anodic discharge behavior of a soluble lithium hydride electrode was studied in the LiCl–KCl eutectic at 380°C, for possible application as a high-energy, low-weight electrode material [A 177].

A pellet cell of the type

$$Ag–LiH/KCl–LiCl/CaCrO_4–Ag$$

gave an open circuit voltage of 2.15 V. Flash currents of 20 A/in.² were obtained; current densities of 2–5 A/in.² were sustained for 4 min to a 1-V end point.

Based on a two-electron oxidation, coulombic efficiencies of 35–60% were computed. The variations of the coulombic efficiency and the large amount of hydrogen gas produced during discharge suggested that two competing reactions were occurring, *i.e.*,

$$LiH \rightarrow Li^+ + H^+ + 2e^-$$
and
$$LiH \rightarrow Li^+ + \tfrac{1}{2}H_2 + 1e^-$$

Positives

Chlorine. The chlorine electrode in LiCl–KCl at 450°C is apparently reversible; in fact, the Cl_2/Cl^- couple on graphite has been used as a reference electrode [A 169].

Many of the problems in adapting this electrode to a battery configuration are much the same as those using a gas electrode in aqueous systems. Apparently, flooding is less of a problem with an electrolyte of LiCl, which, reportedly, does not wet graphite [A 174].

It should be possible to again borrow from high-temperature fuel-cell technology in developing stable compact electrode structures and gas delivery equipment. Chlorine electrodes have been operated as part of H_2/Cl_2 fuel cells. A system has been recently described [A 178] which employs an electrolyte of molten NaCl (42 mol.%)–LiCl at 450–600°C. The electrode was formed out of uncatalyzed, porous carbon; electrodes with 60-μ pores were more effective than those with 40-μ pores. The activation overvoltage was 65 mV at 600°C for a current density of 30 A/dm² (1.94 A/in.²). The electrode did not deteriorate with use for <100 hr. Details of cell construction were not available, except that the electrode spacing was 10 mm.

Oxygen. The experimental data quoted below on oxygen and the oxides of copper, platinum, palladium, nickel, and vanadium in the KCl–LiCl system were taken principally from the work of Laitinen [A 161,166]. These studies consisted of measuring the reversibility of these electrodes in equilibrium with oxide. Varying amounts of lithium or calcium oxide were added to establish the oxide-ion concentration.

Measurements have been made on the O_2/O^{-2} couple in LiCl–KCl (400–500°C) employing a graphite electrode. This oxygen electrode, when dipped into solutions 0.09–0.21M in lithium oxide, was very slow to reach equilibrium at temperatures of 400–500°C. After 6–7 hr, the potential usually had reached a constant value, the final potentials at 450°C being −0.212, −0.328, and −0.407 V against a 1M Pt(II) reference electrode at oxide-ion concentrations of 0.094, 0.167, and 0.21M, respectively. The rate of change of potential with concentration was much greater than predicted by the Nernst equation. Similar behavior was observed at 400 and 500°C. It was concluded that reversible behavior could not be achieved under these conditions. It is possible that carbon takes part in the potential-determining reaction in view of the finding that, at 800°C in KCl–NaCl saturated with CaO, the potential-determining reaction of a carbon electrode is as follows:

$$2Ca^{+2} + CO_2 + 4e^- \rightleftharpoons C + 2CaO$$

Consequently, the drifting potentials and nonequilibrium character in the present study may be due to mixed electrode processes.

Nickel Oxide. The nickel–nickel oxide couple

$$NiO + 2e^- \rightleftharpoons Ni + O^{-2}$$

was also studied. Nickel is known to form NiO as a stable oxide and no oxychloride of this metal is known. The solubility of nickel oxide, calculated from the potentials, was 3.3×10^{-4} moles/liter. A plot of $\log[O^{-2}]$ *versus* E showed a considerable departure from the theoretical slope and, therefore, indicated irreversibility of the system. The E^0_{oxide} value was approximately -1.23 V, which corresponds to solubility of approximately 10^{-3} moles/liter, calculated from the theoretical slope. The experimental points were scattered, and, thus, very little theoretical significance could be attached to the E^0_{oxide} value. The value -1.23 V is slightly more positive than that expected from $E^0_{\text{Ni}^{+2}-\text{Ni}}$ and solubility of NiO estimated polarographically (-1.297 V).

Two reasons may be advanced for the discrepancy between theory and experiment. First, the Ni(II)–Ni system may act irreversibly in the presence of excess oxide ion. This explanation is implausible because of the high exchange current in the Ni(II)–Ni electrode and the moderate solubility of NiO in LiCl–KCl. Second, a higher nickel oxide, *e.g.*, Ni_2O_3, may be formed. In this instance, the measured potentials would be mixed potentials due to the presence of NiO and Ni_2O_3. This explanation appears to be more plausible, particularly because the formation of the latter oxide has been shown to be promoted by the presence of lithium oxide. Evidence for the existence of Ni_2O_3 at temperatures of 660°C and higher in molten lithium sulfate–potassium sulfate eutectic has also been published.

Copper (I) Oxide. The solubilities of Cu_2O, as determined from potential measurements and from a current–voltage curve, were 3.8×10^{-2} and 5.8×10^{-2} moles/liter, respectively. A plot of $\log[O^{-2}]$ *versus* E gave a slope of 66.2 mV and an E^0_{oxide} value of -1.207 V. Corresponding to the electrode reaction

$$Cu_2O + 2e^- \rightleftharpoons 2Cu + O^{-2}$$

the system Cu/Cu_2O, O^{-2} acts as an electrode of the second kind.

Platinum (II) Oxide. A saturated solution of PtO in equilibrium with solid oxide when brought in contact with the platinum foil electrode registered a potential of -0.0546 V *versus* a $3.28 \times 10^{-2}M$

Pt(II) reference. This potential was stable over a long period (4–5 hr). From the potential, the solubility of PtO was calculated to be 3.32×10^{-3} moles/liter.

Palladium (II) Oxide. The solubility of PdO from potential measurements was calculated as 9.4×10^{-3} moles/liter. The plot of log $[O^-]$ *versus* E gave a slope of 74.2 mV and a value of E^0_{oxide} of -0.514 V. The solubility calculated from E^0_{oxide} is 8.1×10^{-3} moles/liter. The attainment of equilibrium in this system was also extremely slow. From the experimental data, however, the electrode system appeared to behave as a reversible electrode of the second kind.

Copper (II) Oxide. This oxide, alone or with Cu_2O, has been widely used as a cathode in thermal batteries. Ideally this material should discharge as follows:

$$CuO + 2e^- \rightarrow Cu + O^{-2}$$

However, because of the existence of a stable intermediate oxide, the reaction could be expected to proceed as follows:

$$2CuO + 2e^- \rightarrow Cu_2O + O^{-2}$$
$$Cu_2O + 2e^- \rightarrow 2Cu + O^{-2}$$
$$CuO + Cu \rightarrow Cu_2O$$

This reaction of CuO and copper has been observed in aqueous electrolytes.

As mentioned above, the Cu_2O/Cu couple is reversible to O^{-2} in the LiCl–KCl electrolyte; the solubility is approximately 4×10^{-2} moles/liter. This solubility can be expected to complicate cell behavior due to "chemical short-circuiting."

A number of test cells of the following type have been built [A 180] and operated at 600°C:

$$Ca/LiCl, KCl, MgO/CuO/C \text{ or } Cu$$

The open circuit voltage was generally between 2.3 and 2.4 V; the peak load voltage was 1.45 V; the peak current was 188 mA/cm²; the energy density was reported as 177 W-hr/lb of active material, *i.e.*, electrodes, grids, electrolyte, and separator.

Cu(II), Cu(0). As mentioned, the Cu^{+2}/Cu^0 couple is reversible, having a high exchange current. Thus, positives containing soluble copper salts would be expected to discharge at high rates. One such system is $CuCl_2$.

The use of $CuCl_2$ as a positive material appears to be complicated by the highly oxidizing nature of this compound [A 190]. For example, gold substrates rapidly dissolved (10 min) upon immersion in the molten $CuCl_2$ + LiCl + KCl. A tantalum wire, 50 mils in diameter, completely dissolved within 15 hr in the melt at 525°C. A tungsten wire (25 mils) partially dissolved. However, pyrolytic graphite, "glossy carbon," and boron carbide were stable. The polarization at 150 mA/cm² was only 0.03 V.

$CuSO_4$ has also been used as a positive [A 180]. At temperatures of the order of 600°C, the sulfate ion is inert as a depolarizer and the half-cell reaction is essentially

$$Cu^{+2} + 2e^- \rightarrow Cu^0$$

However, diffusion of SO_4^{-2} to the negative would result in sulfide formation. Cells of the type

$$Ca/KCl, LiCl, CuSO_4/Cu, C$$

allegedly gave power outputs of 300 W-hr/lb based on active material, or 140 w-hr/lb based on electrodes, grids, electrolyte, and spacer. Similarly, cells made with $CuPO_3$–$LiPO_3$ as the depolarizers and LiCl–KCl as electrolyte also allegedly produced up to 293 W-hr/lb on an active material basis or 222 W-hr/lb on a total cell basis [A 180].

Polarization studies made by the use of a fast interrupter, which permits the viewing of polarization decay within several microseconds, established that very little activation decay was present. Ohmic effects amounted to only 200–300 mV at 100 mA/cm², while the total experimentally observed polarization was about 1 V. Experiments in which the polarization decay was viewed at slow sweep times (>2 sec/cm) indicated that concentration polarization amounted to only 50–100 mV at this current density. In addition, open circuit potentials after several minutes of operation are normally several hundred millivolts below initial open circuit potentials. A significant energy loss through internal shorting was indicated.

Iron Oxides. This material has also been used in thermal batteries [A 162]. The discharge behavior has been observed to vary with the particular oxide employed—Fe_3O_4, α-Fe_2O_3, or γ-Fe_2O_3.

The physical and chemical reasons for differences in behavior have been discussed on the basis of two broad considerations. One of these is the amount of available reducible iron. Thus, a magnetite cell containing only 3.42 mg-ions of Fe(III) polarizes more rapidly than a γ-Fe$_2$O$_3$ cell with 4.96 mg-ions. The addition of conductive carbon enhances behavior at the beginning of a discharge, but voltage drop is very fast, presumably because of nongalvanic reduction of the iron oxide.

The other general consideration is the specific resistance of the oxide. In each instance discussed above, those effects which enhance conductance lead to higher cell voltage and those which increase resistance result in lower terminal voltages, albeit flatter curves result because of decreased discharge current. In this regard, the performance of the magnetite and γ-Fe$_2$O$_3$ spinels is to be compared with that of α-Fe$_2$O$_3$ obtained as such or thermally prepared; samples containing small amounts of silica have a higher resistivity than pure samples, and this is reflected in cell performance; small amounts of titanium ion decrease resistivity, and this is also reflected in cell performance. Unfortunately, this resistance parameter, which is of primary interest, is difficult to measure directly. It is affected by such factors as compacting pressure, particle size, and porosity.

It can be inferred that one may select a cathode reactant to yield electrical characteristics which are particularly desired in a given instance. For example, if a large voltage is wanted with greater energy output for a short-time discharge, then a γ-Fe$_2$O$_3$ or magnetite might be chosen. If a flat discharge curve is more important than large voltage, then an α-Fe$_2$O$_3$ should be used. The voltage can be shifted up or down to some extent by incorporating small amounts of TiO$_2$ or SiO$_2$, and a high peak voltage of short duration may be had by blending the iron oxide with conductive carbon. Behavior which is between that of two separate samples can be obtained by mixing the samples.

Fe(III), Fe(0). The soluble iron cathode is also reversible and has a high exchange current. Cells of the following type have been built employing this material [A 180]:

$$Ca/LiCl, KCl, Fe_2(SO_4)_3/Fe, C$$

FeCl$_3$ has also been used as a positive material [A 190]. Apparently, no difficulty was noted with gold substrates. A cathode mixture of

1 mole $FeCl_3$ in 4.5 moles of LiCl–KCl eutectic showed a polarization of 0.1 V at 150 mA/cm^2 and 450°C.

Sb_2O_3. This compound has also been discharged as a positive plate at 600°C in the KCl–LiCl eutectic. The solubility of Sb_2O_3 was 0.5% in KCl–LiCl at 600°C. Apparently, the half-cell reaction proceeds directly to the metal [A 180]. Chemical analysis of the melt during the discharge indicated the presence of antimony and undischarged Sb_2O_3.

Cells of the type

$$Ca/KCl, LiCl, MgO/Sb_2O_3$$

had open circuit voltages ranging from 2.2 to 2.4 V and energy densities of 110–120 W-hr/lb of active material, *i.e.*, electrodes, grids, electrolyte, and separator.

The cell performance was not particularly sensitive to Sb_2O_3 content over a wide range of cathode mix composition. However, when the Sb_2O_3 content fell to 45% of the cathode cake, a considerable reduction in output occurred. In such mixtures, dilute in Sb_2O_3, additional graphite is needed; it is not possible to depend on a continuous film of antimony bridging the rather distantly separated particles of depolarizer.

V_2O_5. The solubility of V_2O_5 in the KCl–LiCl eutectic is quite extensive; at high concentrations, V_2O_5 reacts with the melt to form chlorine and some type of insoluble reduced vanadium compound.

Chemical analysis [A 181] of the reaction products formed in mixtures of V_2O_5 and molten LiCl–KCl at 380–500°C over the concentration range 5–30 mol.% V_2O_5 indicates decomposition reactions of the following type:

$$V_2O_5 + 6LiCl \rightarrow 2VOCl_3(g) + 3Li_2O$$
$$V_2O_5 + 10LiCl \rightarrow 2VCl_4(g) + Cl_2(g) + 5Li_2O$$
$$V_2O_5 + 2LiCl \rightarrow V_2O_4 + Cl_2(g) + Li_2O$$
$$x\,Li_2O + y\,V_2O_4 + z\,V_2O_5 \rightarrow x\,Li_2O \cdot y\,V_2O_4 \cdot z\,V_2O_5(s)$$

Vanadium oxytrichloride ($VOCl_3$) was volatile only under vacuum. Less than 10% of the vanadium was reduced by self-discharge from V(V) to V(IV). A blue crystalline solid, isolated from the melt at 550°C and analyzed, gave values of $x = 7$, $y = 3$, and $z = 23$. Its powder X-ray pattern was similar to those of the vanadium bronzes.

In a LiCl–KCl eutectic melt at 450°C, V_2O_5 is electrochemically reduced to an insoluble mixed lithium–vanadium oxide at a potential which is a function of V_2O_5 concentration. Chemical analysis of the lithium vanadate bronze which was formed in the potential region 0.00–0.20 V *versus* a Pt(II) (1 M)/Pt reference electrode gave a compound with a stoichiometry of $Li_2O \cdot 2V_2O_4 \cdot 4V_2O_5$. With dynamic non-equilibrium conditions, electrochemical measurements indicated that the equivalent of 0.5–0.7 electrons (n) were required to reduce one V_2O_5 molecule, whereas n values of 0.85 and 0.8 were obtained from equilibrium electrochemical measurements [A 166].

For the Nernst equation which can be derived from the solid $Li_2O \cdot 2V_2O_4 \cdot 4V_2O_5$, n cannot be larger than 0.67 electron per V_2O_5 regardless of the various possible equilibria between V_2O_5 and O^{-2} which can be considered.

One factor which could cause the discrepancy between dynamic and equilibrium values of n is the adsorption of V_2O_5 or vanadate ions into the surface of lithium vanadate crystals to change the surface composition. These crystals are composed of mixed metal oxides which could present active sites for the adsorption of acidic or basic species.

Chromates. Chromates, *e.g.*, K_2CrO_4, have been used as soluble depolarizers in thermal batteries [A 159,163,164,182]. It has been shown [A 183] that K_2CrO_4 in LiCl–KCl is reduced electrochemically at platinum electrodes to give an insoluble lower oxide of chromium, probably containing lithium ion and occluded melt. The situation is even more complex and probably far more involved than previously indicated [A 163], *i.e.*,

$$2x \ CrO_4^{-2}(l) + 3Ni \rightarrow 2x \ O^{-2}(l) + x \ Cr_2O_3 + 3NiO_x$$
$$NiO_x + 2(x - y) e^- \rightarrow NiO_y + (x - y) \ O^{-2}(l)$$

where l indicates material in the electrolyte.

More recently, the following cell has been described:

$$Ca/KCl–LiCl–AgCl–K_2CrO_4/Ag$$

Based on the observed cell performances with AgCl or K_2CrO_4, or both, the following hypothesis was offered as the cathode discharge process. Silver chloride is part of the initial electrolyte and discharges as follows:

$$AgCl + e^- \rightarrow Ag + Cl^-$$

It can be formed by way of the following reaction:

$$2CrO_4^{-2} + 6Ag + 6Cl^- \rightarrow 6AgCl + Cr_2O_3 + 5O^{-2}$$

which replaces the silver ion that was used galvanically. Other chromates which have been employed as depolarizers [A 182] are calcium chromate, basic sodium zinc chromate, and potassium dichromate. With most of these materials, the voltage drifts with time, and so no stable equilibrium potential could be measured. Cells employing these depolarizers have been classified according to the peak voltage generated in cells. The voltages in Table 3-XII were obtained at current density of 110 mA/in.^{2} based on depolarizer area.

TABLE 3-XII

Peak Voltages in KCl–LiCl at Current Density of 110 mA/in.^{2}*

Couple[†]	Peak voltage
Ca/ZnCrO$_4$	3.28
Ca/CaCrO$_4$	2.88
Mg/CrO$_4$	1.96
Ca/SrCrO$_4$	2.62
Mg/SrCrO$_4$	1.84

*Taken from [A 182].

[†] A nickel collector was used at the positive plate.

The particular depolarizer used can have an influence on anode polarization. For example, when CaCrO$_4$ is used, a magnesium anode polarizes much more than in a cell in which Fe$_2$O$_3$ is employed. This is due to the solubility of CaCrO$_4$, which in LiCl–KCl is 7 wt.% at 400°C and 19 wt.% at 600°C [A 264].

Nitrate. Nitrate ion has also been used as a depolarizer in a thermal battery of the following type [A 184]:

$$Mg/LiNO_3, KCl/Ag$$

Best results were obtained with 50% KCl giving an open circuit voltage of 1.5 V; platinum plates were superior to silver. It was originally assumed that silver was oxidized by LiNO$_3$ to give AgCl, Li$_2$O, and oxides of nitrogen. The silver chloride would be discharged electrochemically. However, it has been shown [A 185] that molten nitrate does not react with silver or platinum, but discharges directly to oxide ion plus oxides of nitrogen.

Sulfate Electrolytes. Lithium sulfate and potassium sulfate (80 mol.%) form a eutectic which melts at 535°C. When this melt is electrolyzed with two platinum electrodes at 625°C, the electrode reaction at the positive plate appeared to be [A 155]:

$$SO_4^{-2} \to \tfrac{1}{2} O_2 + SO_3 + 2 e^-$$

Continuous and vigorous gas evolution occurred at the electrode with no observable solid deposits during an electrolysis at a current density of approximately 10 mA/cm². The potential at the electrode at the beginning of electrolysis was about 0.7 V after *IR* correction *versus* the standard reference one molal Ag(I)–Ag(0) electrode and drifted to a stable value of 0.9 V after 15 min.

In the cathode compartment, both sulfite and sulfide were found by chemical analysis, accounting for over 95% of the electricity passed. The cathode process was complicated and probably involved the following total reactions:

$$SO_4^{-2} + 2e^- \to SO_3^{-2} + O^{-2}$$

$$SO_4^{-2} + 6e^- \to S + 4O^{-2}$$

$$SO_4^{-2} + 8e^- \to S^{-2} + 4O^{-2}$$

However, elemental sulfur, if present at all, appeared to be a minor product. The cathode potential at the start was about −1.6 V after *IR* correction *versus* one molal Ag(I)–Ag(0) and drifted to a final stable value of −2.0 V after 35 min.

Table 3-XIII gives the emf series developed; the one molal Ag(I)–Ag(0) electrode was used as the reference electrode.

It is to be noted that the potential span of this electrolyte is much shorter than the alkali chloride melts. The more active metal electrodes cannot be studied conveniently because of chemical reduction of sulfate. For example, nickel was observed to react with the melt to form nickel sulfide. In a melt also containing Na_2SO_4 [A 186], the following metals were also observed to decompose sulfate to form

TABLE 3-XIII

Emf Series in Li_2SO_4-K_2SO_4 Eutectic at 625°C

Electrode	Standard potential
(SO_4^{-2}, O_2)	+0.7
Pd(II)–Pd(0)	+0.541
Rh(III)–Rh(0)	+0.387
Cu(II)–Cu(I)	+0.051
Cu(I)–Cu(0)	−0.201
(SO_4^{-2}, S^{-2})	−1.6

sulfides: magnesium, iron, cadmium, zinc, thallium, and lead. The reaction with magnesium seemed to be faster than those with other metals.

Note that this corrosion reaction does not preclude the use of active anodes in this electrode, but it does limit the life of the battery and the efficiency of the negative plate. However, it is necessary that the sulfide be at least partly soluble in the melt to prevent passivation. A similar situation exists with the aqueous magnesium batteries.

It has been reported [A 180] that at higher temperatures (800°C) anions, such as sulfate and phosphate, may actually serve as depolarizers. A complex discharge scheme has been proposed, involving intermediates such as $S_2O_2^{-2}$, $S_2O_4^{-2}$, and $S_2O_6^{-2}$. A group of test cells involving borates, silicates, sulfates, and phosphates was screened. Initially, Kaolin clay, which had been used as a separator in thermal cells by others, was used in making these cells. However, Kaolin is an aluminosilicate and, thus, is an oxygenated anion acting as a depolarizer. No such problem was experienced with a MgO separator. Testing of Ca/K_2SO_4-Li_2SO_4 cells, where the salt serves as both an electrolyte and depolarizer, resulted in cells which allegedly yielded up to 127 W-hr/lb on an active material basis. The Ca/KPO_3-$LiPO_3$ cell gave 220 W-hr/lb on the same basis.

Secondary Cells

The possible use of molten salt electrolyte cells as secondary batteries has been considered [A 189]. As indicated previously, many couples are quite reversible; the validity of Faraday's law has been demonstrated rigorously for the electrolysis of molten salts. Furthermore, the electrolytes have a high conductivity which should minimize IR losses at high rates of charge and discharge.

The major problem in developing secondary cells using molten salt electrolytes is the method of separation of the anode and cathode, because the various reversible materials show some solubility in the molten salt electrolytes.

Chapter 4

Electrochemically Active Materials: Nonaqueous, Organic Electrolytes

The development of high-energy-density battery systems requires high potential and low equivalent weight plate materials. Metals which fulfill these criteria are the elements listed in the upper left-hand corner of the periodic table, *e.g.*, lithium and magnesium. These materials are sufficiently active to decompose water chemically, which precludes the use of aqueous electrolytes, except possibly for high-rate reserve batteries. One class of compatible electrolyte is the aprotic, organic solvent.

ROLE OF THE SOLVENT

Briefly, the role of the electrolyte may be divided into the following three interrelated parts:

1. Ion transfer through the solution.
2. Mass transport of reactants and products to and from the electrode surface.
3. Participation in the charge transfer process within the double layer.

The basic chemistry involved in charge and material transport through solution has been discussed in Chapter 1. Obviously, a high conductivity is desirable; often, but not always, such electrolytes have a high dielectric constant and a low viscosity. From the increasing data on nonaqueous electrolytes, it has become apparent that explicit account must be taken of the role of the electrolyte in the electrode charge and discharge reactions; the efficiency of discharge is very much dependent on the particular solvent and, in some cases, the solute.

The experimental study of ion–solvent interactions in nonaqueous media has been rather limited. Only a few of the solvents of interest for battery applications have been investigated. Polarographic studies have been made of the behavior of metal ions in acetonitrile (AN). For example, it was found that most metal ions and anions reflect a lower solvation energy in AN than in water. Cuprous and silver ions were the most notable exceptions. As a result, the redox behavior of copper ions was decidedly different from that in water [A 242–244]. The polarographic behavior of metal ions in dimethyl sulfoxide has been investigated [A 245,246], as well as the behavior of alkaline earth metal ions in N, N'-dimethyl formamide [A 247]. It was found that DMF solvates simple metal ions much more strongly than acetonitrile, but less than water.

SPECIFIC ELECTROLYTES

A large number of solute and solvent combinations are theoretically available for use with primary batteries, according to the criterion that the electrolyte is to be inert to direct chemical attack by the electrode materials. The stability requirements are somewhat more stringent in the case of secondary batteries, since the electrolyte must also be stable over the added voltage range required for charging the plates. For example, in the lithium–nitromethane system, the solvent, apparently stable on discharge, may be decomposed on charging the negative plate.

The more widely investigated solvents for battery use are as follows: (1) the cyclic esters, *e.g.*, propylene carbonate and butyrolactone; (2) linear esters, *e.g.*, methyl formate; (3) dimethyl formamide; (4) nitriles, *e.g.*, acetonitrile and n-butylnitrile [P 382]; (5) nitromethane; (6) dimethyl sulfoxide; (7) saturated aliphatic amines [P 389], *e.g.*, propylamine [A 79]; and (8) N-nitrosodimethylamine [A 358]. For simplicity of presentation, the abbreviations listed in Table 4-I will be used in indicating the particular solvent under discussion. The solutes which have been most frequently studied are the following: (1) simple salts, *e.g.*, LiClO$_4$ [P 389] and Mg (ClO$_4$)$_2$ [P 383–388]; (2) Lewis acids alone [P 382], *e.g.*, AlCl$_3$, BF$_3$, and ZrCl$_4$, or combined with an alkali halide [A 75]; (3) complex fluorides, *e.g.*, NaPF$_6$ and KBF$_4$; and (4) complex salts,

TABLE 4-I

List of Abbreviations

Material	Abbreviation
Acetonitrile	AN
Butyrolactone	BL
Dimethyl formamide	DMF
Dimethyl sulfoxide	DMSO
Ethylene carbonate	EC
Ethyl ether	EE
Methyl formate	MF
Nitromethane	NM
Propylene carbonate	PC
Tetrabutyl ammonium iodide	TBAI

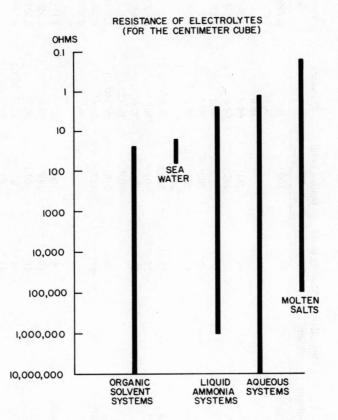

Fig. 4-1. A comparison of electrolytes on the basis of resistance.

TABLE 4-II

Physical Properties of Some Solvents*

Solvent	Molecular weight	Melting point (°C)	Boiling point (°C)	Dielectric constant	Viscosity (cP)	Specific conductivity (Ω-cm)⁻¹	Density (g/cm³)
Water	18	0	100	78^f 78.5^g 78.6^e	0.895^e 1.0^f	$10^{-8\ b}$ $4 \times 10^{-8\ c}$	0.99^f
PC	102	$-49^{c,f}$	241^c 242^f	$64^{c,f}$ 64.4^a 69.0^e	2.5^f 2.53^a 2.55^e	$10^{-7\ b}$ $1.0 \times 10^{-7\ e}$ $2 \times 10^{-7\ e}$	1.19^f 1.20^a
BL	86	-42^g $-43^{c,f}$	202^a $204^{c,f}$ 206^g	$39^{c,f}$ 39.1^a	1.7^f 1.75^a	$10^{-7\ b}$	1.12^f 1.1286^a (16°C) 1.13^a
NM	61	-28.5^g $-28.6^{d,h}$ -29^c	100.8^g $101^{f,h}$ 101.3^d 102^c	35.87^d (30°C) 36^f 37.0^e 39^e (20°C)	0.6^f 0.627^h 0.63^e	$6 \times 10^{-7\ d}$ (18°C)	1.13^f 1.131^d 1.1311^h 1.1354^g (22°C)
EC		36	248	89^f (40°C)	1.9^f (40°C)		1.323^f
DMF	73	$-61^{d,g}$	$153^{d,g}$	26.6^d 36.7^e	0.796^e 0.802^d	$1.206 \times 10^{-6\ e}$ $10^{-7\ b}$	$0.9445^{d,g}$
DMSO	78	6^g 18.45^d	$100^{g\ \dagger}$ 189^d	45^d 46.6^e	1.1^d (27°C) 1.96^e	$2 \times 10^{-7\ e}$ $2 \times 10^{-8\ e}$ $0.87 \times 10^{-7\ e}$	1.100^d (20°C) 1.1014^d (20°C)
AN	41	-41^d -44.9^d -45.7^g	80.1^g $81.6^{d,h}$ 82^c	36.7^e 37.5^d (20°C) 39^e (20°C)	0.325^h 0.344^e	$7 \times 10^{-6\ d}$ (20°C) $10^{-7\ b}$	0.7784^h (23.4°C) 0.7822^h (20°C) 0.783^d (20°C) 0.7856^g (20°C)

* The superscript letters refer to the following references: a, [A ^{73}]; b, [A ^{89}]; c, [A ^{106}]; d, [A ^{108}]; e, [A ^{127}]; f, [A ^{128}]; g, [A ^{148}]; h, [A ^{149}].

† Decomposes.

TABLE 4-III

Table of Conductivities*

Solvent	$LiClO_4$	KPF_6	KCNS	$AlCl_3$	LiCl
Acetone	5×10^{-2} [A 86]				
MF	3.2×10^{-2} [A 86]				
DMF	2.08×10^{-2} [A 127]	2.5×10^{-2} [A 71]	2.08×10^{-2} [A 89]	4.55×10^{-2} [A 89]	8.3×10^{-3} [A 71]
AN	2.85×10^{-2} [A 89]	1.79×10^{-2} [A 89]	2.2×10^{-2} [A 89]	2.28×10^{-2} [A 89]	4.3×10^{-4} [A 89]
BL	1.5×10^{-2} [A 86]	1.3×10^{-2} [A 89] 5.41×10^{-3} [A 89]	1.14×10^{-2} [A 89]	4.66×10^{-3} [A 89]	6.9×10^{-4} [A 89]
PC	4.9×10^{-3} [A 128] 4.76×10^{-3} [A 89]	7.8×10^{-3} [A 71] 7.25×10^{-3} [A 87]	6.02×10^{-3} [A 89]	9.72×10^{-3} [A 127] 7×10^{-3} [A 128]	3.4×10^{-4} [A 71]
NM				1.5×10^{-2} [A 128]	
DMSO			8.8×10^{-3} [A 127]		

* All data are experimental values.

TABLE 4-IV

Specific Conductivities*

Solvent	Solute	Specific conductivity $(\Omega\text{-cm})^{-1}$	Temperature (°C)	Concentration	Reference
NM	$AlCl_3$	1.5×10^{-2}	25	2.2 M	[A 128]
	$AlCl_3$ + LiCl	4.1×10^{-3}	25	3 M	[A 130]
	$AlCl_3$ + LiCl	4.2×10^{-2}	25	1.5 M	[A 128]
	$AlCl_3$ + LiCl	4.70×10^{-2}	25	10 g $AlCl_3$/100 g NM sat. with LiCl	[A 88]
	$AlCl_3$ + LiCl	6.25×10^{-2}	50	3 M	[A 130]
PC	$LiClO_4$	4.76×10^{-3}	24	10 g/100 cc	[A 89]
	$LiClO_4$	4.9×10^{-3}	25	8 g/100 cc	[A 128]
	$AlCl_3$	3.53×10^{-3}	24	5 g/100 cc	[A 89]
	$AlCl_3$	7×10^{-3}	25	15 g/100 cc	[A 128]
	$AlCl_3$	7.5×10^{-3}	25	15 g/100 cc	[A 78]
	$AlCl_3$	9.2×10^{-3}	25	0.6868 M	[A 127]
	$AlCl_3$ + LiCl	6.57×10^{-3}		10 g $AlCl_3$/100 g PC sat. with LiCl	[A 88]
	$AlCl_3$ + LiCl	6.6×10^{-3}	25	12 g/100 cc	[A 78]
	LiCl	3.13×10^{-4}	24	10 g/100 cc	[A 89]
	LiCl	3.4×10^{-4}	25	1 M	[A 71]
	LiCl	2×10^{-7}			
	$NaPF_6$	6.03×10^{-3}	24	0.79 M	[A 129]
	$NaPF_6$	6.8×10^{-3}	24	0.86 M	[A 88]
	KPF_6	3.12×10^{-3}		10 g/100 cc	[A 88]
	KPF_6	3.74×10^{-3}	24	5 g/100 cc	[A 89]
	KPF_6	7.25×10^{-3}	25	1 m	[A 87]
	KPF_6	7.30×10^{-3}	25	1.20 m	[A 87]

TABLE 4-IV (continued)

Specific Conductivities*

Solvent	Salt	Conductivity		Temp	Concentration	Ref.
	KPF_6	7.8	$\times 10^{-3}$	25	1 m	[A 71]
	$LiBF_4$	2.54	$\times 10^{-3}$	24	0.42 M	[A 88]
	$LiBF_4$	4	$\times 10^{-3}$	25	1 m	[A 75]
	KF	2.8	$\times 10^{-5}$	25	1 m	[A 71]
	TBAI	4.5	$\times 10^{-3}$	25	1 m	[A 71]
	TBAI	5.3	$\times 10^{-3}$		1 m	[A 87]
	KCNS	6.02	$\times 10^{-3}$	24	20 g/100 cc	[A 89]
BL	$NaPF_6$	1.11	$\times 10^{-2}$		0.44 M	[A 94]
	$NaPF_6$	1.34	$\times 10^{-2}$	24	1.14 M	[A 88]
	$LiBF_4$	3.40	$\times 10^{-3}$	24	0.25 M	[A 88]
	$LiBF_4$	7.1	$\times 10^{-3}$	25	1 m	[A 75]
	$LiBF_4$	8.3	$\times 10^{-3}$	40	1 m	[A 75]
	$LiBF_4$	9.7	$\times 10^{-3}$	60	1 m	[A 75]
	$LiClO_4$	1.5	$\times 10^{-2}$		10 g/100 cc	[A 86]
	$LiClO_4$	9.7	$\times 10^{-3}$	24	10 g/100 cc	[A 89]
	$Mg(ClO_4)_2$	8.15	$\times 10^{-3}$	24	10 g/100 cc	[A 89]
	KCNS	1.14	$\times 10^{-2}$	24	15 g/100 cc	[A 89]
	$AlCl_3$	4.66	$\times 10^{-3}$	24	5 g/100 cc	[A 89]
	LiCl	6.9	$\times 10^{-4}$	24	10 g/100 cc	[A 89]
Acetone	$LiClO_4$	5	$\times 10^{-2}$			[A 86]
MF	$LiClO_4$	3.2	$\times 10^{-2}$		30 g/100 cc	[A 86]
DMF	LiBr	1.845	$\times 10^{-2}$	25	1.88 M	[A 127]
	$LiClO_4$	1.85	$\times 10^{-2}$	24	10 g/100 cc	[A 89]

TABLE 4-IV (*continued*)

Specific Conductivities*

Solvent	Solute	Specific conductivity $(\Omega\text{-cm})^{-1}$	Temperature (°C)	Concentration	Reference
	LiClO$_4$	2.086×10^{-2}	25	2.88 M	[A 127]
	KPF$_6$	2.37×10^{-2}	27	1 m	[A 87]
	KPF$_6$	2.48×10^{-2}		1.5 m	[A 87]
	KPF$_6$	2.5×10^{-2}	25	1 m	[A 71]
	TBAI	1.0×10^{-2}		1 m	[A 87]
	TBAI	1.1×10^{-2}	25	1 m	[A 71]
	LiCl	8.3×10^{-3}	25	1 m	[A 71]
	NaBF$_4$	1.0×10^{-2}	25	1 m	[A 71]
	NaBF$_4$	2.04×10^{-2}	27	1.01 m	[A 87]
	NaBF$_4$	2.27×10^{-2}	28	1.48 m	[A 87]
	KI	2.2×10^{-2}	25	1 m	[A 71]
	AlCl$_3$	4.55×10^{-2}	24	5 g/100 cc	[A 89]
	KCNS	2.08×10^{-2}	24	20 g/100 cc	[A 89]
DMSO	KCNS	8.8×10^{-3}	25	4.308 M	[A 127]
AN	KCNS	2.2×10^{-2}	24	15 g/100 cc	[A 89]
	AlCl$_3$	2.28×10^{-2}	24	5 g/100 cc	[A 89]
	KPF$_6$	1.79×10^{-2}	24	5 g/100 cc	[A 89]
	LiClO$_4$	2.85×10^{-2}	24	10 g/100 cc	[A 89]
	LiCl	4.3×10^{-4}	24	10 g/100 cc	[A 89]
	Morpholinium PF$_6$	5.69×10^{-2}		1.565 M	[A 301]
Water	NaCl	4.75×10^{-2}		0.5 M	
	KOH	56.5×10^{-2}		3.56 M	
	H$_2$SO$_4$	58.5×10^{-2}		1.55 M	

* All data are experimental values.

e.g., morpholinium hexafluorophosphate and tetraalkyl ammonium salts [A 358].

Some of the physical properties of these solvents are given in Table 4-II. A collection of specific conductivities for a number of solvent–solute systems is given in Tables 4-III and 4-IV. Figure 4-1 shows a comparison of these solvents with inorganic systems.

Cyclic Esters

Of these, PC and BL have received the most attention. As shown in Table 4-IV, the BL solutions generally have higher conductivity.

One of the more widely studied solutes is $AlCl_3$–LiCl. To some extent, the performance of the $AlCl_3$–LiCl electrolytes may be influenced by impurities [A 72]. Solutions prepared by the addition of anhydrous $AlCl_3$ to PC often developed a deep golden brown color. Solutions of better character are obtained by first preparing a diethyl ether solution of aluminum trichloride by bubbling hydrogen chloride through a flask containing diethyl ether and high-purity aluminum metal. The colorless solutions thus obtained are added to appropriate amounts of propylene carbonate from which the diethyl ether was finally removed by vacuum distillation, leaving a nearly colorless solution of aluminum trichloride in propylene carbonate. Polarographic measurements still indicated the presence of an impurity in the solution reducible at potentials more positive than that for lithium reduction. This material could be removed by prolonged electrolysis at 2.5 V. The heat of reaction for $LiCl + AlCl_3 \rightarrow LiAlCl_4$ has been determined as 40 kcal/mole [A 248]. This heat could be expected to induce some decomposition of the electrolyte.

There is some evidence that significant amounts ($>10\%$) of aluminum together with lithium can be plated out from PC–LiCl–$AlCl_3$ solutes together with lithium. However, the analytical technique would not distinguish between metallic aluminum and occluded $AlCl_3$.

The complex fluoride solutes have also been studied in both PC and BL. Apparently, the corrosion problem is more severe and the choice of positive materials is more limited, particularly for secondary batteries. For example, in PC–KPF_6 solutions, copper and silver are apparently oxidized to soluble ions rather than insoluble salt deposits of CuF_2 and AgF. The chemical corrosion properties appear to be dependent on the water content of the solvent. Many common metals are attacked by a $NaPF_6$–PC electrolyte. However, when the electrolyte had less than 50 ppm water, aluminum, silver, and stainless steel were negligibly attacked after 60 days at 165°F [A 75]. The performances of cells containing these electrolytes will be discussed in a later section.

It has been reported [A 249] that mixtures of EC and PC will give conductivities higher than can be obtained with either solvent alone. For example, KPF_6 in 80% EC–20% PC solvent had a conductivity of 1.16×10^{-2} $(\Omega\text{-cm})^{-1}$. This increase is accompanied by a decrease in viscosity and a 35% increase in dielectric constant.

Linear Esters [A 86]

Conductivity–concentration data, obtained for $LiClO_4$ in five such solvents, are summarized in Figs. 4-2 and 4-3. Some physical properties of these esters are listed in Table 4-III. The most conductive solution [3.2×10^{-2} $(\Omega\text{-cm})^{-1}$] was 30 g $LiClO_4$/100 ml of methyl formate. The stability of lithium metal in the esters alone and in solutions with $LiClO_4$ was studied; n-butyl formate appeared completely

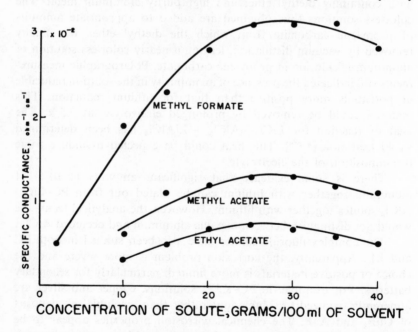

Fig. 4-2. Specific conductance of $LiClO_4$ solutions in methyl formate, methyl acetate, and ethyl acetate [A 86].

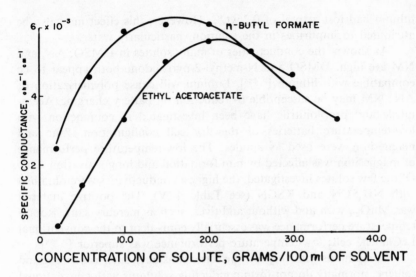

Fig. 4-3. Specific conductance of LiClO$_4$ solutions in n-butyl formate and ethyl acetoacetate [A 86].

inert, and only very slow gassing was observed with methyl formate and methyl acetate. In the case of the last two solvents, a white film could be observed on the lithium surface after five-days exposure. With LiClO$_4$ solute, n-butyl formate gave a rapid reaction, and the reaction with methyl acetate was also accelerated. Such behavior is rather anomalous; an impurity effect is suspected. No difference in behavior was observed in the methyl formate tests between presence and absence of the solute. However, the small gassing still would limit the use of such cells to reserve batteries.

General

A number of cells have been built with an electrolyte containing DMF. Although there is no immediate degradation of the electrolyte, some extended compatibility tests indicate that there may be a problem area for long-term wet-stand operation [A 266]. For example, when lithium wire was placed in DMF, it was noted that the surface was pitted after 48 hr, but had a very bright metallic appearance. After 96 hr, the DMF showed signs of polymerization, but the lithium remained the same. After 7 days, the DMF was almost a gel; the lithium had lost a total of 0.2 wt.%. However, this effect may also be attributed to impurities in the solution, particularly water.

As shown, the conductivities of many solutes in DMSO, AN, and NM are high. DMSO and N-methyl-2-pyrrolidone both appear to be compatible with lithium [A 266]. Lithium will cause polymerization of AN; NM may be susceptible to reduction on battery charging. Acetonitrile and propionitrile have been investigated in conjunction with low-temperature batteries of the dry cell configuration. Zinc and magnesium were used as anodes. The low-temperature performance of magnesium was affected by film formation and long activation time. Of the few solutes investigated, the highest conductivities were obtained with NH_4SCN and $KSCN$ (see Table 4-IV). The positive material was MnO_2, with and without additives, such as mercury salts. Room-temperature performance was essentially equivalent to the conventional LeClanché cell; low-temperature performance was superior [A 238].

Compounds with very low dielectric constants, such as benzene and toluene, normally do not form conducting solutions with conventional salts, but do form electrically conducting systems with materials such as metal alkyls. It is claimed that an addition compound forms between the solvent and the organic aluminum compound, and an increase in conductivity is observed over that of the compound itself. For example, $NaF \cdot 2Al(C_2H_5)_3$ at 45°C has a specific conductivity of 7.87×10^{-3} $(\Omega\text{-cm})^{-1}$; the benzene-solvated complex has a conductivity of 1.11×10^{-2} $(\Omega\text{-cm})^{-1}$ at a ratio of 10 moles of solvent per mole of aluminum complex [P 403]. The addition of an equal volume of toluene to a PC electrolyte containing LiCl and $AlCl_3$ [A 120] results in an increase in conductivity from 6.5×10^{-3} to 7.7×10^{-3} $(\Omega\text{-cm})^{-1}$. The cathodic efficiency for lithium deposition from this solution was 95–100%, while the anodic efficiency was 35–50%. The electrolyte containing ethyl ether, rather than toluene, had an electrical conductivity of 9.8×10^{-3} $(\Omega\text{-cm})^{-1}$ [A 120]. The current efficiencies for deposition and oxidation of lithium are about the same as with the propylene carbonate. Cell tests indicated that the use of the ether–PC electrolyte did give increased capacity.

An ethyl ether solution of LiCl (12 g/100 ml) and $AlCl_3$ (35 g/100 ml) has a conductivity of 2.7×10^{-3} $(\Omega\text{-cm})^{-1}$ [A 120]. However, it was not possible to deposit lithium from such a solution; some aluminum was formed.

A study of the electrochemical stability of 3M $AlCl_3$ and 0.75M LiCl in anhydrous ethyl ether indicated a "decomposition voltage" of 1.95 V [A 359]. "This decomposition seems to be characteristic of the solvent for the following reasons: (1) No products were obtained coming exclusively from the solute, no aluminum or lithium deposition at the cathode, no chlorine evolution at the anode. (2) The products obtained require the decomposition of ether; at the cathode, ethylene (C_3H_4) as well as hydrogen atoms, which form H_2 or HCl, must come from the solvent; at the anode, the brown liquid formed at +1.35 V is probably an organic polymer of low molecular weight. (3) This decomposition voltage seems to be the same with different solutes when the products show a decomposition of the solvent" [A 359].

In acetone–$LiClO_4$ solutions, cathode efficiencies of 90% have been obtained with the CuF_2 paper–carbon electrode [A 85]. However, again the solvent is sufficiently unstable in the presence of lithium, making discharges longer than about the 10-hr rate impractical.

Inspection of the conductance data in Table 4-IV indicates high

TABLE 4-V

Solubility of Perchlorates

Solvent	Moles/100 moles of solvent at 25°C						
	Li	Na	K	Rb	Cs	Mg	Ba
Water	9.5	30.8	0.216	0.130	0.153	8.03	10.6
Ethanol	64.8	5.52	0.0040	0.0024	0.0022	4.94	17.0
Acetone	74.4	24.5	0.065	0.030	0.037	11.1	21.5
Ethyl acetate	78.4	6.93	0.00095	0.00076	0	27.9	29.6

conductivity for aprotic solutions containing large anions or large cations, or both, *e.g.*, morpholinium hexafluorophosphate. It is worth noting the high solubility of $LiClO_4$ in some organic solvents; the data are given in Table 4-V [A 121].

Lithium fluoride is very soluble in butyrolactone in the presence of boron trifluoride, yielding solutions with a conductivity equal to that expected from the metathetical formation of ionized $LiBF_4$, from which lithium can be reversibly deposited.

ELECTRODE POTENTIAL [A 289]

"When standard potentials in nonaqueous solvents are to be determined, it is reasonable to first relate all the electrode potentials (just as in the case of aqueous solutions) to the standard potential of the hydrogen electrode, *i.e.*, for each medium, $E_H^0 \equiv 0$. The numerical values for the potentials are then given by the free energy of the following reaction:

$$\tfrac{1}{2} H_2 + Me^+_{(solv.)} \rightleftharpoons H^+_{(solv.)} + Me$$

There is no doubt that there are large differences between the energy of solvation of the proton in various solvents.

"The series of electrical potentials should shift in such a way that the difference between the standard potentials of one element in two solvents is a measure of the free energy of the reaction $Me^+_{(solv.I)} \rightarrow Me^+_{(solv.II)}$. The resulting energy is the difference between the free energies of solvation or the free energy of the transference of ions. This energy, however, cannot be realized thermodynamically, since together with the positive cation there is also a negative anion transferred from solvent I into solvent II. The energy actually measured is always only the free energy of transference for the whole electrolyte.

"Theoretical considerations on a molecular basis have to be applied in order to separate the energy into single values for the ions. If it were possible to obtain the free energy of transference for just one ion, a common point of departure for both series of electrical potentials would be given.

"It was suggested [A 289] that the standard potential of rubidium be assumed to be equal in all solvents, since the rubidium ion conforms more closely to the concept of an ideal ion. It has the same low free energy of solvation in various solvents, is only slightly polarizable in spite of its large ionic radius, and actually does not tend to form complexes.

"Cesium, which is even more suitable, was not used because the standard potential of cesium in water was not even known at the time. Rubidium, however, can still be used as a reference element, since the standard potentials of rubidium and cesium in all solvents measured so far were found to be almost equal.

"This approach was improved by estimating the small variations of the free energies of solvation of rubidium ions in various solvents.

This was done from the energies of solvation of the alkali halides by means of Born's equation (obtained from the solubilities in the respective solvent)."

The standard potentials for the solvents H_2O, CH_3OH, CH_3CN, $HCOOH$, $HCONH_2$, N_2H_4, and NH_3 were thus determined and are listed in Table 4-VI.

TABLE 4-VI

Series of Electrical Potentials According to Pleskow–Strehlow
with $E^0_{H(H_2O)} \equiv 0$ *

Element	H_2O	CH_3OH	CH_3CN	HCOOH	$HCONH_2$	N_2H_4	NH_3
Ca	−2.92		−3.02	−2.97			−2.94
Rb[†]	−2.92	−2.97[‡]	−3.03	−2.98	−2.92	−2.92	−2.92
K	−2.92		−3.02	−2.89	−2.94	−2.93	−2.97
Na	−2.71	−2.76	−2.73	−2.95		−2.74	−2.84
Li	−2.96	−3.13	−3.09	−3.01		−3.11	−3.23
Ca	−2.76		−2.61	−2.73		−2.82	−2.73
Zn	−0.76	−0.77	−0.60	−0.58	−0.83	−1.32	−1.52
Cd	−0.40	−0.46	−0.33	−0.28	−0.48	−1.01	−1.19
Tl	−0.34	−0.41			−0.41		
Pb	−0.13	−0.23	+0.02	−0.25	−0.26	−0.56	−0.67
H	0	−0.03	+0.14	+0.47	−0.07	−0.91	−0.99
Cu/Cu^{+2}	+0.35	+0.31	−0.24	+0.33	+0.21		−0.56
Cu/Cu^+	+0.52		−0.14			−0.69	−0.58
Hg/Hg^+	+0.80	+0.71		+0.65			
Ag	+0.80	+0.73	+0.37	+0.64		−0.14	−0.16
Hg/Hg^{+2}	+0.85		+0.39				−0.24
Cl	+1.36	+1.09	+0.72	+1.24			+1.04
Br	+1.07	+0.86	+0.61	+0.99			+0.84
I	+0.54	+0.33	+0.21	+0.44			+0.46

* Taken from [A 289].

† E^0_{Rb} values for CH_3OH, CH_3CN, HCOOH, and $HCONH_2$ are corrected to take into account the differences in their energies of solvation in comparison with water; a value of −2.92 V was used for the E^0_{Rb} of N_2H_4 and NH_3. The standard potentials of the halogens are calculated from the solubilities of the alkali halides in the respective solvents.

‡ The standard potential of rubidium in methanol has not been determined. Since the series of electrical potentials in water and methanol are quite similar, the assumption was made that the difference between the standard potentials of sodium and rubidium in both solvents is equal. The comparison of both series of electrical potentials will therefore remain less reliable until the standard potential of rubidium has been measured.

NEGATIVES

Materials

Considerable attention is being given to lithium as a negative material for primary and secondary cells employing aprotic solvents. This metal has a low equivalent weight and a high equilibrium potential. A number of other materials (*e.g.*, magnesium and aluminum) are sufficiently active to be considered. Early work with organic electrolytes was concerned with electroplating these metals. Since these electrolytes are adequate for the mass-transport processes through the bulk of solution and through the electrode–electrolyte interface, they could also be considered for the reverse process, the discharge of the negative material.

Aluminum Anodes. As shown previously, the equivalent weight of aluminum is 8.9, compared to 6.9 for lithium and 32.7 for zinc. Aluminum electrodes, although less active than lithium in the same solvent, are relatively easy to fabricate and handle.

Aluminum can be electroplated from an ethyl pyridinium bromide–66 mol.% $AlCl_3$ solution [A 90]. By adding benzene, xylene, or toluene, a solution is formed with good electrical conductivity at room temperature; *e.g.*, the specific conductivity in toluene is 1.5×10^{-2} (Ω-cm)$^{-1}$ [A 89]. Current efficiencies of 100% are obtained for bright aluminum deposition at current densities up to 20 mA/cm²; above this, the deposits are nonadherent and black. The anodic dissolution efficiency of this bright aluminum rises to 100% at a current density of 20 mA/cm². Inferior results were obtained with benzene and xylene.

Aluminum has been reported to be electrodeposited at 100% current efficiencies in aluminum triethyl–sodium fluoride (Ziegler electrolyte). This solvent is difficult to work with, since aluminum triethyl is pyrophoric and burns explosively on contact with water. This compound may also be too strong a reducing agent to be compatible with most cathode materials. For example, there is some evidence [A 89] that copper fluoride reacts with aluminum triethyl to form copper metal. The addition of toluene to this electrolyte greatly simplifies the handling problems.

Aluminum has been electrodeposited from an $AlCl_3$–n-hexylamine–ether system at current efficiencies of 90% at 10 mA/cm². Aluminum can also be deposited from a solution of LiCl and $AlCl_3$ in ethyl ether at an efficiency approaching 85% [A 120]. However, aluminum is not deposited from a solution of $AlCl_3$ in propylene carbonate. Difficulty was also experienced in attempting to discharge aluminum in electrolytes of: (1) LiBr and acetonitrile, (2) $NaClO_4$ and propylene carbonate, and (3) KCNS and dimethyl sulfoxide. The measured potentials on open circuit were also erratic and irreproducible [A 74].

Magnesium. The equivalent weight of magnesium is 12.2. Current efficiencies of 100% were obtained at current densities less than 2 mA/cm^2 for the electrodeposition and electrodissolution of magnesium in the ethyl magnesium bromide–ether system (Grignard solution). At higher current densities (more negative potentials), reaction of the ethyl groups and ether occurs.

Magnesium was not deposited from a solution of $MgCl_2$ in propylene carbonate; magnesium chloride and ethyl pyridinium chloride do not form conducting solutions. Some success has been achieved in discharging magnesium in $NaClO_4$ in acetonitrile and LiBr in propylene carbonate. An unsuccessful attempt was made to discharge magnesium in an electrolyte of $AlCl_3$ in dimethyl sulfoxide. In these electrolytes, the open circuit potentials obtained with magnesium were not reproducible [A 74]. The discharge of magnesium has also been described for the following nonaqueous solvents containing $Mg(ClO_4)_2$: methanol [P 383]; nitromethane [P 384]; pyridine [P 385]; 2-propanone [P 386]; methylacetate [P 387]; methanamide [P 388]. In these cells, MnO_2 was used as a depolarizer. PbO_2 and AgCl were also satisfactory. Water (0.1%) was added to the methanol as an inhibitor to prevent spontaneous corrosion.

The delayed-action characteristics of magnesium anodes, observed in aqueous systems, apparently also exist in nonaqueous electrolytes. On applying an anodic current to a magnesium electrode, an initial positive potential peak is obtained and is occasionally followed by potential oscillations. Since magnesium readily passivates in aqueous systems due to oxide film formation, it is expected that its performance in organic electrolytes will be particularly sensitive to trace water impurity.

Lithium. Lithium has a low equivalent weight of 6.9 and a high equilibrium potential (Table 4-VI). A number of electrode configurations are being considered in the development of test cells. For example, an attempt was made to electroplate lithium in a pressed composite formed of a lithium salt (LiCl), a conductor, *e.g.*, silver powder, and a plastic binder. Difficulties were experienced because of the hydroscopic character of the lithium salt. More success was had in plating lithium directly from solution onto a metallic conductor. A smooth lithium electrode can also be made by withdrawing a 40-mesh silver-plated nickel screen from molten lithium metal (m.p., 186°C). However, this configuration lacks adherence [A 78].

Lithium electrodes have been prepared by mechanically forming lithium sheet. For example, lithium rod is rolled in an argon-atmosphere glove box to a ribbon 10–20 mil thick and 2 in. wide. A nickel expanded-metal sheet is sandwiched between two pieces of lithium ribbon, and the composite is again rolled. The lithium sheets are bonded together through the screen by the rolling pressure [A 78].

Pasted electrodes [A 91] have also been formed using a dispersion of lithium ($>20 \mu$ particle size) in mineral oil. A uniform layer is spread on both sides of a nickel screen (225 holes/in.^{2}) and pressed at 2 tons/in^2. The electrodes are then washed with hexane to remove the oil. Suitable adhesion is obtained even without a binder. A pasted electrode is formed from methocel (2% in DMSO), lithium, and a conductive material, such as nickel powder ($>10 \mu$). The resulting electrode has a capacity of 0.66 A-hr/g [A 92]. Methocel (MC), a carboxylmethyl cellulose binder, is insoluble in propylene carbonate and soluble in dimethyl sulfoxide. One method of limiting contact between the electrodes and the atmosphere has been to heat-seal the plates and separators within polyethylene envelopes. Electrolyte is introduced with a hypodermic syringe [A 85].

Tests were run on the discharge of a lithium negative, originally in the form of rolled lithium sheet. Approximately 90% utilization was achieved with polarization levels below 200 mV at a current density of 1.55 mA/cm^2 in an electrolyte of $NaPF_6 - PC$ [A 88]. A discharge curve for a lithium anode in 0.46 M LiBr in PC is shown in Fig. 4-4 [A 119]. Some deterioration in performance was noted with time. Coulombic discharge efficiencies of 87–100% were noted at a current density of 4 mA/cm^2 for lithium anodes in LiBr–PC and LiBr–DMSO [A 119].

The electrodeposition of lithium has been studied from a propylene carbonate–LiF–$LiPF_6$ solution and from aluminum chloride–lithium chloride–propylene carbonate. In the latter system, a black deposit was formed which was poorly adherent to the copper base metal. By bagging the electrode with dacron cloth to retain the electrodeposited

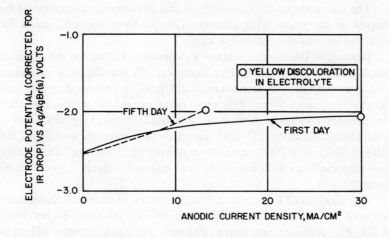

Fig. 4-4. Polarization of lithium anode in 0.46M LiBr–PC [A 119].

material, current efficiencies approaching 100% have been measured for lithium production at 70–80 mA/cm². This efficiency was determined by measuring the total amount of lithium deposited and did not distinguish between lithium and LiOH. During extended periods of electrolysis, no $AlCl_3$ reacted with the electrodeposited lithium or was reduced at the negative [A 88,89].

Lithium can also be deposited with high efficiency from propylene carbonate, in the absence of $AlCl_3$. Efficiency data are given in Table 4-VII [A 120].

TABLE 4-VII

Lithium Deposition in Propylene Carbonate*

Solute	Concentration	Specific conductivity $(\Omega\text{-cm})^{-1}$	Lithium efficiency (%)	
			Cathodic	Anodic
$LiClO_4$	0.64 M	1.9×10^{-3}	95–100	84
LiBr	0.44 M	2×10^{-3}	94–100	80
$LiAlCl_4$	0.63 M	6×10^{-3}	95–100	50

* Taken from [A 120].

The electrodeposition onto silver and platinum is complicated by alloying of the metal with lithium. This problem was not noted for substrates of nickel or stainless steel.

Generally, the solvent exerted a pronounced effect on the physical appearance of the deposit. For example, lithium deposits produced from nitromethane solutions were light gray in color, adherent, and dendritic in form. When deposited from propylene carbonate, they were loosely adherent, dark gray, and amorphous. Deposits from a mixed electrolyte of 1:1 PC–NM were also very adherent and deposited with an efficiency of 100% at a current density of 13 mA/cm². Depositing from nitromethane solution alone was efficient only at higher solute concentrations.

The addition of Rhodamine B sodium salt and disodium fluorescein improved the adherence of the plate in PC solution. Plates from LiClO$_4$–PC were very adherent. Pressure produces a more adherent deposit, as in the case with present AgO/Zn systems. However, electrode porosity must be provided to ensure adequate electrolyte at the interface.

The electrolysis of a solution of tetramethyl ammonium chloride–LiCl–propylene carbonate or butyrolactone system yielded no lithium. In the butyrolactone solution, a light blue color formed which disappeared when air was admitted to the solution. The system AlF$_3$–LiF is a poor conductor in propylene carbonate.

Effects of Impurities

Lithium. Much of the work which has been reported [A 72–75] has been on solvents containing 50–200 ppm (0.02%) water. These trace quantities appear to exert only a minor effect on the discharge of lithium. Larger amounts of water may passivate lithium by forming a coat of LiOH [A 74,75].

The specific influence of water on the discharge characteristics of lithium has been studied with an electrolyte of KSCN in PC. The change in anode potential with increasing water content was negligible, both on open circuit and on load. Even addition of ~5% water to the anode compartment scarcely altered the polarization on load [A 119].

The problems in charging lithium electrodes are more serious; attempts to charge lithium in propylene carbonate containing 200 ppm water were relatively unsuccessful [A 72]. Deposition of lithium from a PC–LiF–LiPF$_6$ electrolyte containing 100 ppm water (0.1 g/liter) was achieved with a current efficiency of 50% at 10–30 mA/cm². Solutions carefully purified or pre-electrolyzed for extended times gave 100% current efficiencies at 25 mA/cm² [A 89].

In the presence of water, the cathodic reactions *could* be as follows:

$$Li(H_2O)^+ + e^- \rightarrow LiOH + \tfrac{1}{2}H_2$$

or

$$Li(H_2O)^+ + Li^+ + 2e^- \rightarrow Li_2O + H_2$$

Alternatively, lithium may be first deposited and then react with water to form the hydroxide. Thus, charge is consumed, but is not available for anodic discharge. Little is known about the solubility of LiOH or Li_2O in nonaqueous solvents.

In defense of the cathodic reactions given above, experimental results [A 84] indicate that water is not reduced independent of and prior to the reduction of lithium. Furthermore, a potential plateau is observed at values lower than the lithium deposition potential only when the concentration of added water is more than four times the concentration of lithium initially present. This suggests that the complexing of water with lithium ions occurs and that the water so complexed is not reducible independent of the reduction of lithium. Essentially the same results are obtained from $LiClO_4$ solutions containing $AlCl_3$, when it would be expected that the water would be scavenged by formation of $Al(OH)_3$ and HCl. Lithium should have a high hydration energy as a result of its smaller ionic radius; *e.g.*, the heat of solution of LiCl in water is 8.5 kcal. However, the heat of hydration of $AlCl_3$ is 78 kcal, so that the preferential reaction of water with $AlCl_3$ would be expected. It has been observed that large amounts of water will liberate HCl [A 84].

A second problem with the formation of an oxide or hydroxide is that the film may be electrically insulating. The situation becomes analogous to that observed for charging a silver positive in aqueous KOH, where a nonconductive Ag_2O is formed first, which then must be further oxidized to conductive AgO. High efficiencies are observed only at low charging rates.

Magnesium Electrodes. The initial open circuit potentials of magnesium electrode systems are generally below the theoretical value. In one set of experiments, charging current was applied to the cell systems to reduce the magnesium surface and to remove moisture electrochemically. The results obtained are shown in Table 4-VIII [A 87]. No significant corrosion of the magnesium was observed.

TABLE 4-VIII

Highest Values Obtained for Open Circuit Voltage*

System number	Cell system	Voltage (V)		
		Initial	After charging	Theoretical
1	Mg/KPF$_6$–PC/CuF$_2$–Cu	1.36	1.44	2.92
2	Mg/NaBF$_4$–PC/CuF$_2$–Cu	1.31	3.0	2.92
3	Mg/KPF$_6$–PC/NiF$_2$–Ni	1.24	3.14	2.21
4	Mg/NaBF$_4$–PC/NiF$_2$–Ni	1.24	2.9	2.21
5	Mg/KPF$_6$–DMF/NiF$_2$–Ni	1.20	3.26	2.21
6	Mg/NaBF$_4$–DMF/NiF$_2$–Ni	1.22	2.42	2.21

* Taken from [A 87].

It has been shown that these large differences are due to a more active negative. The voltages, after charging, do drift with time. For example, system 2 decays to 2.30 V in 10 min; system 3 to 2.7 V in 22 min; system 4 decays to 1.85 V in 10 min. Some of these abnormally high open circuit potentials, after charging, may have been due to the presence of foreign species on the anode, e.g., sodium or potassium deposited from the electrolyte. The slow decay may have been due to reaction of magnesium with water to form Mg(OH)$_2$.

The system Mg/KPF$_6$–PC/NiF$_2$–Ni was selected for a study of the effect of moisture on cell potential. The KPF$_6$–PC solution was exposed to the moist atmosphere for two days; it was found that the high open circuit potential was never achieved and that measured open circuit potentials were unstable [A 87].

In some cases, water has been added to a nonaqueous electrolyte as an inhibitor for corrosion of a magnesium anode. For example, when magnesium is immersed in anhydrous methanol, spontaneous corrosion of the magnesium occurs. This can be inhibited by adding a small quantity of water to the methanol. An amount of water equal to 0.1% of the weight of the methanol is effective for this purpose. The water is consumed in the reaction by which the inhibiting effect is produced [P 383].

These effects are consistent with the explanation that magnesium and the trace water present in those solvents react to form an impervious film of oxide–hydroxide. The protection which can be afforded the metal, of course, depends on the solubility of the film in the electrolyte. Judging from the information presented above, these films tend to be quite stable. The pronounced influence of surface films on anodic discharge in aqueous systems is discussed in Chapter 5. Little information is available on this general subject on nonaqueous systems, with or without high impurity contents.

Purity Requirements

How pure must these solvents be and what is the maximum tolerable quantity of water? The following calculation establishes the order of magnitude involved.

Consider a 1 cm^2 (10^{16} Å^2) surface of *smooth* lithium exposed to electrolyte. The radius of a lithium atom is 1.56 Å, so that approximately

$$\frac{10^{16}}{\pi(1.56)^2} = 1.35 \times 10^{15} \text{ atoms}$$

of lithium are exposed to electrolyte, or

$$\frac{1.35 \times 10^{15}}{6.023 \times 10^{23}} = 2.26 \times 10^{-9} \text{ moles}$$

Thus, 2.26×10^{-9} moles of H_2O would be required to form a monolayer of LiOH on the surface. Consider an electrolyte volume (density = 1) of 1 cm^3 about the lithium surface. This represents

$$\frac{2.26 \times 10^{-9} \text{ moles}}{10^{-3} \text{ cm}^3} = 2.26 \times 10^{-6} \text{ moles/liter}$$

On a weight basis, this becomes

$$18(2.26 \times 10^{-6}) = 3.06 \times 10^{-5} \text{ g/liter} = 4.06 \times 10^{-2} \text{ ppm}$$

This represents the concentration of water sufficient to form a monolayer of LiOH on a smooth surface of 1 cm^2 of lithium exposed to 1 cm^3 of electrolyte.

Reflux distillation of butyrolactone and propylene carbonate can reduce the water content to 20 ppm; the use of molecular sieves can reduce the water level to 10 ppm, although inorganic salts may be introduced. In one experiment [A 88], dehydration with molecular sieve followed by reflux distillation at reduced pressure reduced the water content of butyrolactone from 2500 to 10 ppm, of propylene carbonate from 395 to 11 ppm, and of dimethyl sulfoxide from 75 to 12 ppm.

A similar calculation can be made on the quantity of oxygen which will form a layer of Li_2O on a slab of smooth lithium. The general problem being considered is how much oxygen would be gettered by handling lithium electrodes during cell assembly. A 1 cm² surface of lithium involves approximately 2.26×10^{-9} moles of lithium atoms.

From the stoichiometry,

$$4Li + O_2 \rightarrow 2Li_2O$$

it is seen that 5.6×10^{-10} moles of oxygen are sufficient to form a monolayer. This amount of gas is

$$(5.6 \times 10^{-10})(22.4) = 1.25 \times 10^{-8} \text{ liters}$$

Thus, one liter of gas containing 0.01 ppm O_2 is sufficient to generate a monolayer of lithium oxide.

With care, an inert gas, such as argon, can be purified to a level of 1 ppm O_2. Advantage can also be taken of rate factors minimizing the time of exposure of active metals. Preferably, the surface of the metal is kept wet so that oxygen must diffuse through a liquid layer to attack the metal. This, however, assumes a water-free, de-aerated liquid film. The rate of diffusion of oxygen through a liquid film is

$$\sim D \frac{\Delta C}{\Delta x} \text{ moles/cm}^2\text{-sec}$$

where D is the diffusion coefficient ($\sim 4 \times 10^{-5}$ cm²/sec), ΔC is the concentration difference across the film, and Δx is the thickness of the liquid film ($\sim 10^{-2}$ in. or $\sim 2.5 \times 10^{-2}$ cm). C_{O_2} (g) is usually $\sim 1.5 \times 10^{-6}$ moles/cm for 1 atm O_2. For 1 ppm in the liquid phase (if Henry's law is assumed), C_{O_2} (g) is $\sim 1.5 \times 10^{-12}$. Then, according to the above equation, the rate of oxygen diffusion is

$$4 \times 10^{-5} \left(\frac{1.5 \times 10^{-12}}{2.5 \times 10^{-2}} \right) \cong 2.2 \times 10^{-15} \text{ moles/cm}^2\text{-sec}$$

Since 1 cm² of perfectly smooth lithium is $\sim 2.2 \times 10^{-9}$ moles and 1 mole of oxygen corresponds to 2 moles of lithium, therefore a monolayer of Li_2O would result in

$$\sim \tfrac{1}{2} \left(\frac{2.2 \times 10^{-15}}{2.2 \times 10^{-9}} \right) \cong 5 \times 10^5 \text{ sec}$$

So the effect of an inert film of liquid is certainly very good. Even if oxygen were 1000 ppm, we would still have 5×10^2 sec for a monolayer of Li_2O, but only ~ 5 sec if we wished to avoid more than 1% contamination of the surface.

POSITIVES

The halogens are active oxidants, fluorine being the most active. The problems encountered in dealing with gaseous depolarizers and in handling fluorine are formidable. It is possible to use transition-metal

TABLE 4-IX

High-Energy-Density Electrode Couples at 25°C*

Reaction		Cell potential (E^0)	Energy density (W-hr/lb)
$2Li + F_2$	$\rightarrow 2LiF$	6.05	2740
$2Li + CuCl_2$	$\rightarrow 2LiCl + Cu$	3.06	505
$2Li + CuF_2$	$\rightarrow 2LiF + Cu$	3.55	754
$2Li + NiF_2$	$\rightarrow 2LiF + Ni$	2.83	620
$2Li + NiCl_2$	$\rightarrow 2LiCl + Ni$	2.57	437
$3Li + CoF_3$	$\rightarrow 3LiF + Co$	3.64	965
$2Li + CoF_2$	$\rightarrow 2LiF + Co$	2.88	633
$2Li + CuO$	$\rightarrow Li_2O + Cu$	2.25	587
$2Li + NiO$	$\rightarrow Li_2O + Ni$	1.79	492
$Mg + CuF_2$	$\rightarrow MgF_2 + Cu$	2.92	566
$Mg + NiF_2$	$\rightarrow MgF_2 + Ni$	2.21	445
$3Mg + 2CoF_3$	$\rightarrow 3MgF_2 + 2Co$	2.89	691
$Mg + CuO$	$\rightarrow MgO + Cu$	2.30	538
$Mg + NiO$	$\rightarrow MgO + Ni$	1.83	451
$Mg + AgO$	$\rightarrow MgO + Ag$	2.98	491
$Ca + CuF_2$	$\rightarrow CaF_2 + Cu$	3.51	604
$Ca + NiF_2$	$\rightarrow CaF_2 + Ni$	2.82	501
$Ca + CuO$	$\rightarrow CaO + Cu$	2.47	503
$Li + AgCl$	$\rightarrow LiCl + Ag$	2.84	231

* Taken from [A 71].

fluorides, such as CuF_2; because of the stability of LiF ($\Delta G_{\text{formation}}$ = −139.6 kcal), a high cell potential results. Calculated cell potentials and energy densities for some couples are shown in Table 4-IX. Of these systems, those employing the halides of copper, nickel, and silver are receiving the most attention, and these will be discussed in some detail below.

There is a possible second-order advantage in the use of metal salt positives, which arises from the relative densities of the salt *versus* the metal discharge product. For example, NiF_2 has a density of 4.6 g/cm^3, and bulk nickel has a density of 8.9 g/cm^3. Thus, on discharge, a volume contraction could be expected, which should result in additional pore formation within the electrode structure [A 89]. This effect, while desirable in the discharge of primary batteries, may lead to problems in recharging secondary cells.

Most of the positive materials listed in Table 4-IX [A 71] are electrical insulators. Hence, some method must be provided to ensure electrical conductivity. The fact that the discharge product is an electrical conductor should aid in withdrawing power from a primary cell, since conductive filaments are built up through the electrode. Some of the details of construction are also given below.

Materials

Specific plate materials are discussed below. The performances of cells employing these materials are discussed in Chapter 8.

CuF₂ [P 389]. The direct synthesis of this compound from the elements is apparently difficult, a mixture of materials being obtained. A working electrode can be formed by mixing this product with copper powder, graphite, and polyethylene (dissolved in hot toluene) and pasting the wet mixture on a 40-mesh screen. After drying, the pasted screens are pressed at 10 tons/in². Electrodes prepared as described, but without the polyethylene, were generally characterized by poor adhesion and poor utilization. Binders of acrylic resin would decompose CuF_2 [A 89].

If the reaction product is soluble, fluoride-containing electrolytes would be desirable for reversibility of the cathode. Solutes such as sodium hexafluorophosphate and sodium fluoroborate in PC and BL have been used. DMSO has also been considered; however, 0.03 moles/liter of CuF_2 were soluble [A 94].

Another electrode mix consisted of 60% CuF_2 and 40% silver flake sintered at 200°C onto a silver screen. At low current densities (1 mA/in.²) in $NaPF_6$–PC, the utilization of CuF_2 was 57%. A somewhat more successful cell was obtained with a mix of 70% CuF_2 and 30% silver flake; 72% of the available CuF_2 was discharged to a 2.0-V end point at 1 mA/in.² [A 94].

Electrodes have been prepared by mixing paper pulp and carbon with the active materials (in the ratio 1:1:12) and pressing the mixture onto an expanded metal support in a steel mold at pressures of 500 to 10,000 psi [A 85]. $LiClO_4$ added to the cathode formulation gave improved efficiency, but was discontinued due to the possible hazard of blending dry $LiClO_4$ with carbon. The thickness of the cathode did not appear to affect discharge efficiency, nor did the capacity appear to depend on the discharge rate over the 40–250-hr range, *i.e.*, slow discharges [A 86]. Electrodes of this type have been discharged in a BL–Mg $(ClO_4)_2$ electrolyte at 0.5 mA/cm² [A 85] for an efficiency of 55%.

A bobbin construction is also possible. One part of the cathode material was blended with three parts of graphite and 2 g of the mix poured around a $\frac{1}{4}$-in. carbon rod to make electrodes 1 in. high and 0.5 in. in diameter.

The solubility of CuF_2 in some solvents was determined by chemical analysis [A 119]. The results are given in Table 4-X. These solubilities need not prohibit the use of CuF_2 as a positive plate. However, it would be necessary to prevent migration of the soluble copper to the negative where it could plate out, discharging lithium.

It is possible that these solubility data reflect the presence of impurity within the electrolyte.

TABLE 4-X

Solubility of CuF_2

Solution	Solubility (%)
0.4 M KSCN/PC	0.046
0.4 M KSCN/AN	0.112
0.4 M LiBr/AN	0.252
0.33 M NaClO₄/DMSO	0.047

CuF_2 is insoluble in $AlCl_3$–ethyl pyridinium bromide–toluene solution; however, CuF is soluble and copper metal will deposit on less noble negatives.

Electrochemical discharge efficiencies of a CuF_2 cathode were in the range 50–70% with the methyl formate electrolyte, while those with butyrolactone were generally in the range 20–30%. No particular care was used to remove water impurity.

$CuCl_2$. This depolarizer has received some attention; however, its use is complicated by a significant solubility in solvents such as PC and DMF [A 84,250]. In a study of the following cell [A 250]:

$$Li/LiCl, DMF/CuCl_2$$

it was found that the large self-discharge was due to the dissolution of $CuCl_2$. Soluble copper salts were found in the separator; copper metal was found at the negative. The measured solubilities of CuCl (5×10^{-2} moles/liter) and $CuCl_2$ (2.5 moles/liter) in DMF were sufficient to account for the observed self-discharge of the cell. Efforts to reduce the solubility of copper chlorides in DMF by the use of the common-ion effect, mixed solvents, and complexing anions were not effective. This solubility restricts the use of this system to reserve battery applications.

CoF_3. Electrodes of this material were prepared in a manner similar to that for CuF_2. However, the CoF_3 was inactive electrochemically, and there were indications of decomposition during sintering.

$NiCl_2$. This material is insoluble in the PC–$AlCl_3$–LiCl system. The electrochemical formation of $NiCl_2$ is very much dependent on the current density used. At low current density, *i.e.*, 5 mA/in.^{2}, 80% efficiency is observed; at 20 mA/in.^{2}, the efficiency drops to 20%.

One approach to electrode construction involves impregnating a porous nickel sheet with $NiCl_2 \cdot 6H_2O$, followed by removal of the water [A 91]. NiF_2 electrodes were prepared in a similar manner [A 92] by impregnating a porous nickel plaque with nickel hydroxide which was reacted with HF. The discharge characteristics were poor.

The halide-impregnated plaques appeared to benefit from a formation cycle. Utilization factors as high as 80% have been obtained. However, there is a possibility that the nickel substrate may have participated, which would obscure the utilization data.

During discharge, a precipitate which contains 2.2% aluminum and some nickel may also form at the electrode surface and in the electrolyte [A 92]. There is some evidence for interaction between the $NiCl_2$ electrode and the electrolyte, which apparently results in a chloride deficiency within the electrolyte. This "chlorine-starved condition" results in a small increase in electrolyte resistivity and a significant polarization at the positive. Treatment of the electrolyte with chlorine significantly decreases this problem [A 266].

This chlorination treatment also has an additional effect on the physical properties of the electrolyte. As mentioned previously, the addition of $AlCl_3$ to PC induces a deep brown color. The rate at which this color appears often depends on the method of addition. The treatment of these solutions with chlorine reduces this coloration to a pale yellow [A 266]. The chemical reactions involved are not known.

AgO. Silver oxide has also been employed as a positive material for nonaqueous solvents, such as PC and AN [A 119]; the fate of the oxygen on discharge is unclear. One electrode formulation was a four-component mixture consisting of equal parts (by weight) of silver shavings, AgO powder, carbon-black beads, and sulfur powder (which seems to facilitate the reduction of AgO). This electrode, in 0.46 M LiBr/AN, showed 0.2-V polarization under a load of 10 mA/cm²; in 0.46 M LiBr/PC, a polarization of 0.9 V was obtained at the same load. This behavior implies considerable participation of the electrolyte in the discharge of AgO.

Exposure of the LiBr/PC electrolyte to moist air had little effect on the open circuit potentials. However, the increased water content ($\sim 1\%$) did appear beneficial with respect to polarization behavior. In one test, the polarization at 10 mA/cm² was reduced by 0.4 V. Similar results were found with an electrolyte of 0.46 M LiBr in DMSO. At 10 mA/cm², the polarization was reduced 1 V by using an electrolyte with 1.7% water [A 251].

Carbon Black. Oxidized carbon black could also be discharged in 0.46 M LiBr/PC at a load of 2 mA/cm². Apparently, the functional groups present on the carbon can be electrochemically active. Some caution is, therefore, needed when using these materials as "inert" supports.

AgCl. The theoretical equilibrium potential of the Li/AgCl couple is 2.84 V and Q_0 is 231 W-hr/lb. It is to be noted that the theoretical capacity of the Zn/AgO system is 234 W-hr/lb. Silver chloride, like $NiCl_2$, is insoluble in a $PC-AlCl_3-LiCl$ electrolyte. Silver was found to form silver chloride at 100% efficiency at current densities up to 72 mA/cm^2. The use of $LiClO_4$ as the electrolyte solute results in a soluble silver oxidation product. However, in the presence of sufficient $AlCl_3$, insoluble AgCl is formed as an adherent coating. Electrodes can also be formed as a mixture of AgCl and powdered silver fused onto a silver screen. However, fused silver chloride tends to produce a nonconducting layer and localized spots of high current

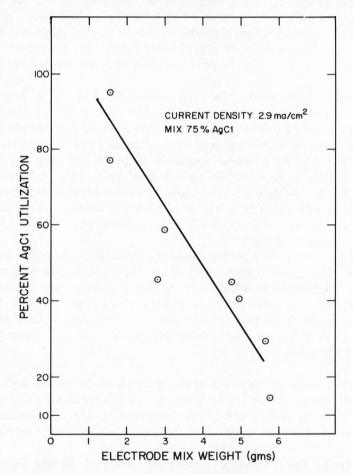

Fig. 4-5. Effect of mix weight on silver chloride utilization in the cells $Li/PC-AlCl_3-LiCl/AgCl-Ag$ [A 120].

density. More successful structures were obtained by first forming the plate from metal, followed by electrical charging. This has been done with pasted powder and with silver sheet [A 93], using binders of silver oxide or graphite. The utilization efficiency of silver chloride increases with the use of thin electrodes. In some cases, current efficiencies of 70% have been noted for this electrode containing 0.25–0.50 g AgCl/in². Efficiencies of 40–60% are more common. The use of graphite in the mix appears to be advantageous for the thick electrodes.

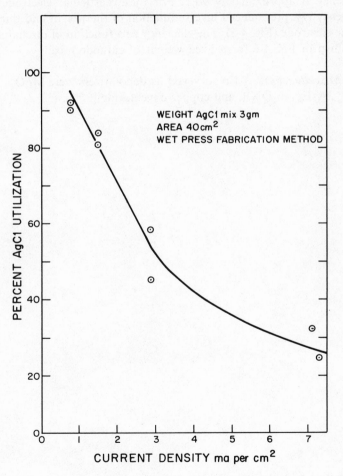

Fig. 4-6. Effect of current density on silver chloride utilization in the cell Li/PC–AlCl₃–LiCl/AgCl–Ag [A 120].

One procedure for preparation is as follows: 50% precipitated AgCl, 10% graphite, and 40% Ag$_2$O wetted with 2% polyvinyl alcohol and water, pasted on silver expanded metal, and sintered in air at 400°C. Electrode thickness is 0.030 in., and it has 70% porosity.

A variation of this wet-paste technique [A 120] involves pressing this wet screen at 500 psi, followed by drying it in an oven at 120°C. Silver flake is used in place of AgO. At current densities below 0.5 mA/cm^2, the efficiency of use of AgCl is 100%. At 2.9 mA/cm^2, the efficiency is approximately 90%. For these particular electrodes, the efficiency is a pronounced inverse function of the weight and thickness of the electrode (Fig. 4-5). The efficiency as a function of discharge rate is shown in Fig. 4-6 for a fixed weight of cathode mix.

Miscellaneous. Also surveyed as depolarizers were MnO$_2$, AgF, AgF$_2$, AsF$_3$, m-DNB, and copper oxychloride [A 358].

Chapter 5

Electrochemical Efficiency K_1

The efficiency with which the active material initially loaded into a battery plate is actually consumed by the cell reaction is determined in part by the extent of side reactions, *e.g.*, corrosion or solubility. Many high-energy plate materials are thermodynamically unstable in the particular battery electrolytes presently employed, leading to corrosion reactions which can eventually self-discharge the device. Techniques for minimizing corrosion by modification of electrolyte or the electrode surface are described below.

Kinetic and mass-transport factors are adversely influenced by a decrease in operating temperature. However, there are a number of important commercial and military applications for batteries which involve a low-temperature ambient. This problem is also discussed in the present chapter.

SELF-DISCHARGE PROCESSES

The corrosion rate of a metal often depends on the nature of a thin surface film formed on the metal. In the case of metals such as titanium and aluminum, these surface films serving as barriers are so effective that rapid corrosion proceeds for only a short time. When transport or migration through the film becomes the controlling factor, the rate of further corrosion is reduced to a small value. Although passivating films are beneficial in the sense described above, these films also limit the discharge rate of the plate. In effect, the maximum rate of discharge is determined by the stability, porosity, and electrical conductivity of the surface film.

The dependences of corrosion rate and discharge efficiency on external parameters, such as current density, electrolyte composition, anode composition, and additives are discussed below in terms of oxide films formed on an electrode surface operating with an aqueous electrolyte. Although much of the information is given in terms of zinc, magnesium, and aluminum, the *concepts* developed are applicable to almost all electrodes. Thus, similar effects are expected in nonaqueous systems, although here the film may be an insoluble fluoride or chloride, rather than an oxide or hydroxide.

TABLE 5-I

Electrode Potentials of Magnesium, Aluminum, and Zinc in Typical Electrolytes*
(Potentials Referred to a Standard Hydrogen Electrode)

Metal	3% NaCl		0.1 N HCl		0.1 N HNO$_3$		0.1 N NaOH	
	1 min	Rubbing	1 min	Rubbing	1 min	Rubbing	1 min	Rubbing
Mg	1.418	1.500	1.622	1.596	1.270	1.220	1.086	1.484
Al	0.577	1.221	0.493	0.916	0.320	0.804	1.403	1.386
Zn	0.772	0.818	0.769	0.752	0.688	0.643	1.126	1.123

* Taken from [A 58].

A qualitative illustration of the effect of surface films is given by the electrode potentials of magnesium and aluminum shown in Table 5-I. Measurements were made by first determining the electrode potentials of the metals in a conventional manner and then during a continuous cleaning of the surface by means of a carborundum rod, which removed much of the original protective film. The table shows that when protective films are formed, they have a substantial effect on the electrode potential. It should also be noted that the formation of a protecting film depends on the composition of the electrolyte. Thus, a film is formed on aluminum immersed in solutions of NaCl, HCl, and HNO$_3$, but the film dissolves in NaOH. On the other hand, protecting films are formed on magnesium in NaOH, but not in NaCl, HCl, or HNO$_3$.

Negative Electrodes

Self-discharge of negative plates is most pronounced in the case of active metals, *i.e.*, metals whose electrode potentials are substantially negative to the reversible hydrogen potential in the electrolyte of interest. The discussion below deals primarily with zinc, aluminum, and magnesium, since these are typical active metals.

As shown in Chapter 2, the potential of the Zn/Zn^{+2} couple is

$$Zn^{+2} + 2e^- \rightarrow Zn \qquad\qquad E^0 = -0.763 \text{ V}$$

Rapid, spontaneous dissolution of zinc is expected and observed in acid solutions. In alkali, zinc oxidizes according to

$$Zn + 2H_2O \rightarrow Zn(OH)_2 + 2H^+ + 2e^- \qquad E^0 = -1.245 \text{ V}$$

The hydrogen half-cell potential in unit activity base is -0.823 V. Thus, zinc is unstable to hydrogen evolution by approximately 0.4 V, and the fact that it can be used at all is due to a high overvoltage for the discharge reaction. Hydrogen evolution, though slow, complicates the construction of sealed cells. In strongly alkaline solutions, stability may be afforded the zinc anodes by the addition of zincate salts to the electrolyte in substantially saturating concentration.

The influence of pH on the corrosion rate of zinc in mixtures of NaCl with NaOH and HCl [A 48] is shown in Fig. 5-1. It is seen that

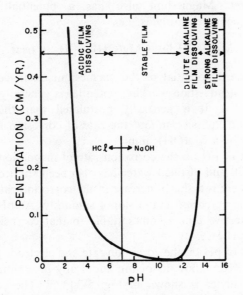

Fig. 5-1. Effect of pH on the corrosion of zinc by NaOH and HCl [A 50].

zinc is rapidly corroded by solutions whose pH is less than 6 or greater than 12.5 owing to the high solubility of the corrosion product formed. In solutions having pH values from 6 to 12.5, the low corrosion rates are attributed to the presence of precipitates which are adherent and, consequently, protective.

In the LeClanché cell, with an NH_4Cl–$ZnCl_2$ electrolyte, corrosion of zinc becomes a problem only on long-time, intermittent use. In this electrolyte, which has an initial pH of 4–5 depending on the concentration of the NH_4Cl and $ZnCl_2$, the corrosion of zinc is expressed by the following overall reaction:

$$Zn + 2NH_4Cl \rightarrow ZnCl_2 + 2NH_3 + H_2$$

In alkaline dry cells, the corrosion of zinc is faster if the reversible potential of the metal is shifted to more negative potentials by complexing zinc, e.g., by NH_3 to form $Zn(NH_3)_4^{+2}$.

Aluminum, like zinc, is amphoteric, its oxide being soluble in both acid and alkali; consequently, it is attacked by electrolytes of very high or very low pH. Curves analogous to that of zinc have been obtained for aluminum, with a minimum near pH 7 [A 49].

Magnesium. Magnesium also has a potential sufficient to decompose aqueous electrolytes, *i.e.*,

$$Mg + 2H_2O \rightarrow Mg(OH)_2 + H_2 + heat$$

The standard potential of the magnesium electrode is -2.43 V. However, the steady-state working potential is generally of the order of -1.5 V [A 192]. It is generally postulated that the magnesium electrode is in a "passive state;" the rate of corrosion and discharge are controlled by a $Mg(OH)_2$ film.

The effect of pH on the corrosion rate of magnesium in solutions of NaOH, HCl, and distilled water has also been studied [A 51]. From pH 3 down, there is a sharp increase in the corrosion rate, while from pH 3 to 11 the corrosion curve slopes smoothly. At pH's of 11 and higher, a protective film becomes stable so that corrosion effectively ceases above pH 11.5.

The effect of pH on the corrosion rate and electrode potential of a magnesium AZ10A alloy dissolving anodically in various magnesium bromide electrolytes is shown in Fig. 5-2 [A 52]. The corrosion rate decreases approximately linearly with increasing electrolyte pH over

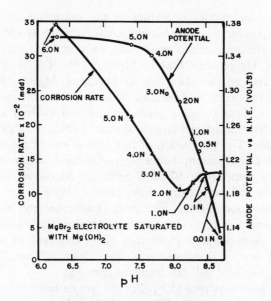

Fig. 5-2. Effect of pH on the corrosion rate and anode potential of a magnesium AZ10A alloy dissolving anodically in various $MgBr_2$ electrolytes saturated with $Mg(OH)_2$ (at a current density of 2.0 mA/cm²) [A 52].

the pH range 6.3–8.1 and, at pH's greater than 8.1, approaches a constant value, corresponding to an anode efficiency of 63%. In the weakly acidic solutions, the potential varies very slightly with pH, while in the basic pH range the potential falls rapidly with increasing pH and decreasing bromide-ion concentration.

These results have been explained in terms of a $Mg(OH)_2$ film on the anode surface, which is more easily penetrable at low pH values because the OH^- ion concentration is too low to maintain an effective oxide film. Thus, at low pH values, high corrosion rates and high anode potentials are observed. At high pH values, the OH^- ion concentration precipitates Mg^{+2} ions so close to the anode as to stifle the anodic reaction [A 53]. This results in lower potentials, a decrease in local corrosion, and excessive polarization of the anode. In an intermediate pH range, precipitation of $Mg(OH)_2$ occurs at a sufficient distance from the active anode areas to permit the anodes to function normally.

The effects of anions on surface oxides are often substantial. Because of their penetrating power, chlorides are not acceptable with magnesium anodes; a magnesium bromide electrolyte is used instead.

Magnesium anodes operating in the inorganic salt electrolytes (e.g., $MgCl_2$) achieve coulombic efficiencies of only 50–65% due to this corrosion. The magnesium anode efficiency in $Mg(ClO_4)$ is approximately 15% higher than that in $MgBr_2$ [A 12]. The shelf life in Mg $(ClO_4)_2$ is also greater than that in $MgCl_2$ or $MgBr_2$. An activated shelf life of one week is obtainable with the proper magnesium alloys. Nevertheless, there is still a great deal of heat evolved at high rates (27 mA/cm²) with the perchlorate system, sufficient to dry out a test cell in 24 min.. The disposition of this heat is one of the major design problems for magnesium batteries, particularly at the higher drain rates. Energy outputs of 50–55 W-hr/lb are obtainable at the 30 min to 30 hr rate at room temperature. Over 40 W-hr/lb are obtained at −50°C with good battery insulation. Unfortunately, the same cell configuration is not applicable both at room temperature and at −50°C. It can be demonstrated [A 13] that the electrolyte of an insulated magnesium battery designed for a 10-hr drain will reach the boiling point after 3 hr and will fail from overheating.

It has been reported [A 11] that the anode behavior with organic salt electrolytes is markedly different. With such electrolytes, coulombic efficiencies of 90–96% were measured without an appreciable increase in the usual potential loss. However, the resistivities of the organic salt electrolytes, e.g., magnesium acetate, are of the order of 2×10^{-2} $(\Omega\text{-cm})^{-1}$ or five times that of inorganic salt electrolytes. The use of mixed electrolytes showed only limited success.

With inorganic electrolytes, the apparent wasteful corrosion rate of magnesium increases directly with increasing anodic current—the "negative difference" effect. The effect of current density on corrosion rate is shown in Fig. 5-3.

Some of these results can also be explained in terms of conventional theory of corrosion, specifically in terms of the "difference effect" and the "chunk effect." The latter refers to accelerated corrosion around grain boundaries and the loss of material in large pieces as opposed to uniform corrosion across the electrode surface. Indeed, on discharge, magnesium anodes tend to become perforated. It is, therefore, desirable to cement the sheet to a supporting conductive substrate and, thus, prevent premature cell failure via this mechanism [P 189].

A number of things can be done to decrease the severity of this corrosion problem. First of all, trace quantities of iron, cobalt, and nickel in the electrolyte must be avoided [P 46], since these greatly

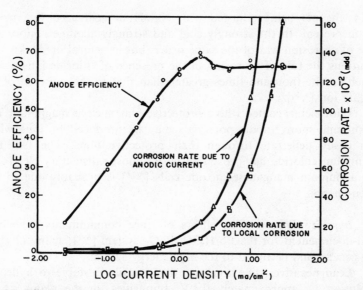

Fig. 5-3. Effect of current density on the local corrosion rate and anode efficiency of a magnesium AZ10A alloy dissolving in 2 N MgBr$_2$ electrolyte saturated with Mg(OH)$_2$ [A 52].

increase the rate of corrosion. It is also possible to include a corrosion inhibitor in the electrolyte. The most common of these is chromate ion [P115,146–149].

The corrosion stability of magnesium can be increased by alloy formation with aluminum, zinc, and manganese [P 144,147]. The role of the aluminum is generally considered to be the prevention of the adherence of insoluble magnesium salts on the electrode surface [P 123].

Aluminum. The behavior of aluminum in aqueous solutions is quite similar to that of magnesium. Similar effects of anions on film stability are found.

For aluminum in a solution containing 0.001 M K$_2$CrO$_4$ and various potassium salts, it was found that small ions show the highest penetration of the oxide film, the order of penetration being Cl$^-$, Br$^-$, I$^-$, F$^-$, SO$_4^{-2}$, NO$_3^-$, and PO$_4^{-2}$ [A 54]. These differences may be connected not only with size, but also with the solubility of the corresponding aluminum salts. What this means in terms of corrosion is seen by measurements [A 55] with aluminum in solutions of NaCl, Na$_2$SO$_4$, and NaNO$_3$. The corrosion rate was determined over the pH range

0–14, the pH being adjusted with NaOH or acid according to the anion present. In the strongly acid and strongly alkaline regions, the rate of corrosion was of the same order, but in neutral or weakly acid solutions the rate of corrosion in the presence of chloride ions was as much as one thousand times greater than that in the presence of the sulfate ion [A 55].

With metals coated with a protective film, such as magnesium and aluminum, more anodic potentials may be produced by ions which can easily penetrate through their protective films. The fact that aluminum chloride has proved to be the most satisfactory electrolyte for aluminum–manganese dioxide cells [A 56] is probably due to this phenomenon.

Impurity Effects. The effects of trace contaminants have been well-documented for the $Pb/H_2SO_4/PbO_2$ system [A 1]; many of these are general and pertinent to the high-energy systems.

Lead negatives, like zinc, are theoretically unstable to hydrogen evolution by approximately 0.4 V. Impurities, in the plates or the electrolyte, which decrease the hydrogen overvoltage, are detrimental to battery performance. For example, traces of platinum (0.0001%) accumulating at a lead negative will produce violent gassing and disruption of the plate. Even 0.00002% platinum will double the rate of self-discharge. Copper, bismuth, silver, arsenic, and antimony impurities have less effect.

Acetic acid, derived from organic materials within the cell, is detrimental to the lead–acid cell. Soluble lead acetate is formed; lead ion then reacts with sulfuric acid to form the insoluble sulfate, regenerating acetate ion. This then constitutes a mechanism for removal of lead from the plates.

Iron and manganese in the electrolyte are also detrimental. For example, the following reaction can proceed at the positive:

$$PbO_2 + 2FeSO_4 + 2H_2SO_4 \rightarrow PbSO_4 + Fe_2(SO_4)_3 + 2H_2O$$

Ferric ion can migrate to the negative plate where the following reaction takes place:

$$Pb + Fe_2(SO_4)_3 \rightarrow PbSO_4 + 2FeSO_4$$

Manganese causes similar reactions. Insoluble MnO_2 is formed; nevertheless, some $HMnO_4$ does remain in solution and has a destructive oxidizing action on organic matter, *e.g.*, separators. The concentration involved is a few hundred thousandths of a percent.

Another type of self-discharge discussed here is the air oxidation of the negative. This is particularly important with the dry-charge battery and reserve type. A number of materials have been added to the negative to minimize this problem, *e.g.*, aryl hydroxy benzoic acids [P 94] and 1-hydroxy-2-naphthoic acid [P 100]. The use of inert atmospheres, *e.g.*, nitrogen or argon, during plate formation is also helpful.

Prevention—Negatives

Inorganic Inhibitors. In some batteries, as much as 50% of the anode material may be consumed by corrosion. To prevent or lessen such corrosion, many expedients have been suggested and tried. Zinc anodes can be protected by mercury, by mercury salts, and by chromic salts in the electrolyte. The main effect of mercury is to raise the hydrogen óverpotential and, thus, decrease corrosion. Mercury must be uniformly distributed at the anode surface. Nonuniform amalgamation can lead to a corrosion reaction between the zinc and the amalgamated zinc. Because of its embrittling effect on conventional anodic materials, only small amounts of mercury can be used.

The use of the double chromate $BaK_2(CrO_4)_2$ rather than $BaCrO_4$ has been reported to improve corrosion resistance of dry cells [P 335]. The same effect has been reported for zinc negative [P 347] and for magnesium anodes in water–methanol electrolytes [P 348,349]. A comparison of the single and double chromate protection is shown in Table 5-II. Identical magnesium cells, except for the corrosion

TABLE 5-II

Comparison of Corrosion Inhibitors

Condition	Time to end point (hr)		Capacity maintenance (%)	
	Lot 1 Sodium chromate	Lot 2 Barium–potassium chromate	Lot 1 Sodium chromate	Lot 2 Barium–potassium chromate
Fresh	271	293		
Three months	164	211	60	72
Twelve months	109	138	40	47

inhibitors, were discharged under identical conditions. The time to the 1.33-V end point is tabulated.

Chromate inhibitors often prove unsatisfactory, as they are affected by the nature of the cell paste and often lose their effectiveness during cell operation.

Organic Inhibitors. Quinoline forms a coat on zinc which will inhibit corrosion. However, it forms a varnish-like film of high electrical resistance. This problem reportedly is not found with two derivatives—8-nitroquinoline and 8-chloroquinoline—and these compounds have been suggested as corrosion inhibitors [P 350]. It is reported that these compounds are also effective at 160°F, *i.e.*, in a temperature region where amalgamating does not inhibit gas evolution.

A film of cramolin reportedly reduced the gas evolution of an AZ 21Mg/MgBr system from 400 cc/800 hr to 0.1 cc/800 hr [A 223].

Cationic organic compounds, though often quite effective in suppressing corrosion, tend to form films on the anode which have high resistance and also since they react with other cell components. Nonionic, surface-active compounds made by the addition of ethylenic oxide compounds to hydroxyl-bearing compounds inhibit corrosion by raising hydrogen overvoltage at the anode without adversely affecting normal cell performance. One such compound is p-hydroxy-diphenylpolyethylene oxide (PPPG), which can be used either alone [P 299] or in conjunction with amalgamation [P 298]. An example of the improvements produced by this compound is shown in Table 5-III [P 298].

The addition (0.1–10 vol.%) of an organo siloxane (silicone oil) has also been reported to suppress corrosion, improve general perform-

TABLE 5-III

Improvement Produced by Use of PPPG in Suppressing Corrosion*

Depolarizer	Inhibitor	Initial		After 18 months	
		Volts	Amperes	Volts	Amperes
MnO$_2$ (electrolytic)	Hg	1.58	9.3	1.45	2.2
	Hg + PPPG	1.59	7.9	1.49	4.6
	PPPG	1.57	7.7	1.47	4.4

* Taken from [P 298].

ance, and increase charge retention [P 292,406,407] in the lead–acid cell. This material can be added directly to the paste, or a coating can be formed by dipping the plates. Similar results are reported for alkaline–zinc cells. Apparently, the electrochemical behavior of the electrodes is not impaired, *i.e.*, the potential difference and the short-circuit current between pairs of electrodes remains approximately unchanged. This permits the use of more corrosive electrolytes with better electrical conduction. Other advantages of this technique are: (1) the corrosion rate of magnesium is reduced to $\frac{1}{10}$ or $\frac{1}{20}$ in a neutral electrolyte, so that its shelf life is similar to that of zinc; (2) the self-discharge of the LeClanché cell is minimized; and (3) the shelf life of the Reuben cell is increased from 2 to 10 years [P 20]. The true effectiveness of silicone oil on magnesium electrodes has been questioned recently [A 361]. Fluorinated wetting agents have also been used to prevent self-discharge of the lead–acid cell [P 98].

Corrosion can also be minimized by alloy formation. For example, the addition of arsenic (0.15–1%) inhibits corrosion of lead plates [P 185–187]. Salts of tellurium, selenium, and germanium will plate and coat the negative, suppressing hydrogen evolution [P 98]. Evolution of hydrogen by iron can be inhibited by adding $<10\%$ cadmium or antimony [P 102]. Similar improvements in the corrosion of magnesium, aluminum, and titanium by alloy formation have been discussed.

As mentioned, some corrosion of negatives is attributable to trace metals in the electrolyte. These can be removed from solution by adding a complexing agent, *e.g.*, EDTA [P 351] or N, N-bis-(hydroxy-ethyl)-glycine [P 414].

It is, of course, desirable to use pure materials in the initial construction of battery plates. For example [A 224], it has been found that many commercial cadmium and nickel salts do not have the desired purity for use in commercial Ni/Cd batteries. It has been more effective to start with pure bulk metal; the acids necessary to prepare the salts can be obtained in adequate purity.

Other Corrosion-Prevention Techniques. It is reported that the solubility and migration of ZnO_2^{-2} in alkaline AgO/Zn cells can be controlled by the concurrent use of pressure and semipermeable membranes [P 381]. This subject will be discussed in more detail below. Further improvements are produced by adding salts of amphoteric metals to the electrolyte to decrease and control the hydroxide-ion concentration and, hence, suppress the formation of zincate and gassing.

Examples of such materials are aluminates, molybdates, vanadates, and arsenates. One such electrolyte is 40% KOH saturated with aluminate (6.7 wt.%) [P 295,351].

It has been reported that deterioration of the Mg/CuCl battery is significantly slowed down if 5–50% polar solvent is mixed with the aqueous electrolyte. For example, it is possible to activate a cuprous chloride battery with a solution of aqueous glycerin and then defer usage of the battery for as much as 6–12 hr without any substantial sacrifice of the ultimate power or current delivery of the battery.

Positive Electrodes

Corrosion of positive electrodes can also occur by reaction with electrolyte to yield oxygen and a reduction product of the oxide. This mechanism may occur with positives having potentials above the reversible oxygen potential. The rate of the corrosion reaction of AgO in alkaline electrolytes was investigated by a sensitive microvolumetric method [A 36]. The reaction rate increases with hydroxyl-ion concentration and is catalyzed by light. The decomposition proceeds at a rate of about 16% in one year at 30°C and 49% in one year at 45°C. Little gas was evolved by AgO if the temperature of the material remained at or below 25°C [A 225].

Direct dissolution represents another form of self-discharge particularly common to positives. Unlike corrosion, there is no valence change involved. A number of examples have already been given, e.g., m-DNB in aqueous solution and $CuCl_2$ aqueous and organic solutions.

Of particular importance is the solubility of silver oxide electrodes in alkaline solution, which is a factor determining battery performance and life. Dissolved silver ions diffuse across and around separators or membranes and plate out as metallic silver at the negative electrode [A 44]. If the battery is on open circuit, it suffers an equivalent self-discharge by conversion of a corresponding amount of, for example, zinc to zinc oxide or hydroxide.

The solubility of Ag_2O and AgO in alkaline solution on open circuit was studied with a polarographic technique, using a rotating platinum electrode [A 36]. Only monovalent silver species could be detected in solutions which had been in intimate contact with AgO powder over prolonged periods of time. This is not surprising, in view of the thermodynamic instability of AgO noted previously.

Quantitative measurements of the solubility of Ag_2O in KOH solutions ranging from 1–14 moles/liter were carried out utilizing a potentiometric titration method. The solubility has a maximum at about 6 N KOH, where it reaches a value of $4.8 \times 10^{-4}N$ (Fig. 5-4). The presence of zincate in the electrolyte was shown to have only a small effect on the solubility of silver oxides [A 36].

It has been assumed that in alkaline solution Ag_2O produces a soluble anion in the following manner [A 226]:

$$Ag_2O + 2OH^- \rightarrow 2AgO^- + H_2O$$

However, the postulated AgO^- species has been questioned; there is some evidence that in alkaline solution the ions of monovalent silver form a trinuclear complex [A 227]:

$$3Ag_2O + 2OH^- + H_2O \rightarrow 2[Ag_3O(OH)_2]^-$$

The solubility of AgO in alkaline solution nearly coincides with the solubility of Ag_2O [A 228]. It is unlikely that divalent silver would be found in solution, since either of the following reactions would take place:

$$AgO \text{ (aq.)} + Ag \rightarrow Ag_2O$$

$$2AgO \text{ (aq.)} \rightarrow Ag_2O + \tfrac{1}{2}O_2$$

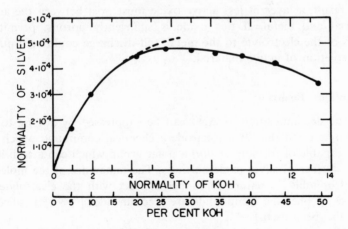

Fig. 5-4. Solubility of Ag_2O in KOH electrolyte as determined by potentiometric titration [A 35].

A study [A 43] was made of the electrical factors which control the amount of silver which is introduced into the electrolyte during the operation of a secondary silver–alkaline battery. It was shown that silver is introduced into the electrolyte during all phases of operation of a silver–alkaline battery. The quantity of silver in solution increases on charging at low current densities or by operation at the Ag/Ag_2O voltage level. The migration of soluble silver was due primarily to diffusion.

The study of this system is complicated by the tendency of soluble silver to be reduced to colloidal silver. Care must be taken to filter out such material before taking measurements [A 354].

There is another self-discharge mechanism involving the auto-decomposition of active materials themselves. For example, on a positive plate containing PbO_2 on lead, the following reaction is possible:

$$PbO_2 + Pb + 2H_2SO_4 \rightarrow 2PbSO_4 + 2H_2O$$

This effect is pronounced with Planté plates, which consist of thin layers of oxide on lead. In general terms, this reaction is due to the existence of three stable oxidation states, a situation not unique to the lead system. The same effect has more recently been observed with silver oxide cathodes according to the following reaction:

$$AgO + Ag \rightarrow Ag_2O$$

As a result, a layer of less active oxide must exist between the active material and metal. If this oxide is sufficiently porous, permitting access of the electrolyte to the metal, self-discharge could continue to the depletion of active material.

Prevention—Positives

The decomposition of AgO can be suppressed by adding to the electrolyte or to the silver electrode a chemical compound which has (or is capable of forming in situ) a polar group which can attach itself to the surface of silver oxide particles. The remainder of the molecule is hydrophobic in nature to hinder contact with the electrolyte. A number of phenolic compounds are recommended [A 20; P 26]; silicones have also been used [A 407].

Alternatively, it has been suggested [P 101] that 0.1–5% Pb be added to the silver positive. Supposedly, a coating of silver plumbate Ag_2PbO_3

is formed, which reduces gassing by a factor of 4, decreases the resistance of AgO, and utilizes close to 100% of the divalent silver plateau. It is believed that the plumbate, a good conductor, coats the silver oxides and is not reduced until after the silver oxides.

Separator Membranes

One method of reducing the transport of soluble material across the electrolyte is by placing a semipermeable separator membrane between the electrodes. It is desirable that the electrolyte conductivity not be decreased substantially while the diffusion of the soluble plate material is prevented. To some extent these requirements are mutually exclusive. The successful development of secondary AgO/Zn batteries has been due, in great part, to the use of suitable membrane separators. For example, a cell without separator will usually fail on the first recharge, whereas a cell with one turn of 1-mil cellophane around the positive will last for several cycles. A cell with six or eight turns will last up to several hundred cycles, depending on cycling conditions [A 296].

The following discussion of separators is taken principally from the work of Weiss [A 234] and Cooper and Fleischer [A 296].

Requirements. In addition to retarding transfer of soluble active material, the general requirements for spacer membranes are as follows:

1. Strong mechanical separation between positive and negative plates.
2. Effective prevention of migration of particles between plates of opposite polarity.
3. Minimum resistance to the flow of electrolyte. This problem is particularly important for space batteries which can be subjected to high acceleration.
4. Good conductivity in the electrolyte.
5. The separator should be easily wetted by the electrolyte. It should absorb and retain a maximum of electrolyte, so that there is always enough electrolyte in contact with both plates to ensure an uninterrupted flow of current.
6. The separator should fulfill all these functions over extended periods of time, both under cycling conditions and in prolonged storage. The material must therefore resist, as long as possible, the

effects of the battery environment; *e.g.*, this entails resistance to the combined effects of hydrolysis in the presence of strong alkali and of oxidation in the presence of the highly oxidative AgO and Ag_2O.

7. Any degradation products of the separator material should not interfere with the proper functioning of the battery.

8. Separators should have sufficient mechanical strength so that they are not damaged in assembling the battery and can withstand reasonable handling of the battery.

9. Separators should also be dimensionally stable, though some degree of swelling due to electrolyte absorption is unavoidable and can be compensated by appropriate cell design.

Experimental techniques for measuring these parameters are described in the work of Cooper and Fleischer [A 296].

Tortuosity. Requirements 2, 3, and 4 involve the porosity and pore structure of the membrane. Pore diameter and porosity calculations are generally based on the assumption that the pores form a system of parallel cylinders, going straight through the matrix, at right angles to its surfaces. Such pores would provide the shortest pathway for ionic migration; their length would be uniform and equal to the measured thickness of the membrane. This is, however, an idealized picture. In reality, the pores are not straight, parallel, and perpendicular to the surface, but form a complex pattern of interconnected, randomly bent capillaries of varying lengths and diameters. The length of the capillaries is greater than the thickness of the membrane. Two membranes may, therefore, have the same average pore cross-sectional area and identical thickness, but may differ widely in mean path lengths. This will result in different electrical resistivities.

The ratio by which the effective length of the pores exceeds the straight line (*i.e.*, the thickness of the membrane) is termed the tortuosity factor:

$$T = \frac{\text{mean effective capillary length}}{\text{thickness of the membrane}}$$

It is the factor by which the separator thickness must be multiplied to arrive at the actual mean path length which the ions must travel in diffusing through the separator. The actual path length will, of course, affect the rate at which ions can travel through the membrane and will, therefore, influence the conductivity.

The tortuosity factor can be calculated by means of the following equation [A 235]:

$$T = \frac{1 + V_p}{1 - V_p} \tag{1}$$

where V_p is the volume fraction of the membrane occupied by the polymer network. To use this equation, the weight and the specific gravity of the dry separator material and the volume swelling of the membrane are required.

From the determinations of electrical resistance and pore size, it is possible to calculate the tortuosity factor by another equation:

$$T = \left(k \frac{R_0 A P}{L}\right)^{\frac{1}{2}} \tag{2}$$

where k is the specific conductivity of the electrolyte [for 31% KOH, $k = 0.66\ (\Omega\text{-cm})^{-1}$]; R_0 is the measured resistance of the separator in the electrolyte; A is the cross-sectional area; P is the porosity; and L is the apparent path length, *i.e.*, the thickness of the separator. The term $R_0 A/L$ represents the specific resistivity ρ of the separator. By the appropriate substitutions, we obtain

$$T = (0.66 \cdot \rho \cdot P)^{1/2} = 0.8123\ (\rho \cdot P)^{1/2} \tag{3}$$

The values for the porosity P can be calculated from the following relation:

$$P = \frac{B}{L} \tag{4}$$

where B is the total pore volume per unit surface area and L is again the thickness of the separator.

The results obtained by the two methods are in good agreement. The tortuosity factor of cellophane PUDO-300 was found to be 2.8 by use of equation (1), while a value of 2.6 was obtained with equation (2). Tortuosity data have been calculated for various separators. Cellulosic materials, cellophane as well as sausage casing, generally have tortuosity factors within a fairly narrow range, from 1.9 to 2.6; *i.e.*, their effective path lengths are about 2–2.5 times the thickness of the sheet. These values represent the lowest range of tortuosity factors, coupled with a high degree of porosity (in the swollen state) and low electrical resistivity. Permion 600 also has the high porosity typical of cellulosic materials, but its tortuosity factor of 5.5 is more than double the value of unmodified cellophane, presumably due to the grafted sidechain; its electrical resistance is about four times as high as that of the unmodified cellophane.

Plastic separators, the PMA membrane ($T = 3.2$–5.4), and Permion 300 ($T = 7.5$) have considerably higher tortuosity factors than the cellophanes. Their resistivity is also much higher, although their porosity is not too different from those of the cellulosic materials. Only the microporous polyethylene membrane Mipor 13 CN and Acropor WA have resistivities which are close to those of the cellulosic separators; in both cases, the average pore size is large, while the tortuosity factors are in the range of the cellulosic materials, *i.e.*, in the low range.

Low tortuosity factors, *e.g.*, in the neighborhood of $T = 2$, are evidently desirable in a separator because low electrical resistance appears to depend on a relatively short path.

On this basis, Mipor 13 CN and Acropor WA should be very desirable separator materials; both have large pore diameters and low tortuosity factors. Unfortunately, they are deficient in another respect; they are not efficient barriers for dissolved silver.

Inhibition of Silver Migration [A 234]. Apparently, the inhibition of silver migration can involve chemical processes as well as the physical effects associated with soluble zinc.

Experiments with tagged silver (Ag^{110}) showed the following: (1) Silver ions diffuse through a porous material which is reasonably stable to oxidation by "Ag_2O;" and (2) there is no measurable diffusion through a material which tends to react with dissolved "Ag_2O."

Cellophane is typical of materials that are readily oxidized. It reacts with "Ag_2O" and there is virtually no silver diffusion through a cellophane membrane up to its capacity to react with "Ag_2O." In a diffusion experiment with Ag^{110} and one layer of cellophane PUDO-300, there was no measurable radioactivity in the untagged cell compartment after 100 hr, at which time there was still some Ag^{110} left in the tagged compartment. If cellophane is impregnated with an effective antioxidant, such as hydroquinone or p-phenylenediamine, its reaction rate with "Ag_2O" is markedly reduced and its permeability to diffusion by dissolved silver species is increased. Plastic membranes which had proved resistant to alkaline permanganate oxidation were also found to react slowly with "Ag_2O," but the dissolved silver ions diffused rapidly through the membranes. These materials, therefore, could not be considered effective separators.

One method of inhibiting this degradation involves forming, upon the cellulose, a thin (0.0002-in.) coating of a polystyrene butadiene latex, in which is imbedded particles of a cation ion exchange rising, e.g., sulfanated polystyrene [P 419]. These materials are reported to be stable in the caustic electrolyte and are sufficiently flexible to prevent cracking due to swelling of the cellulose. A significant reduction in silver penetration was noted. The specific resistivity of the film was approximately 50 Ω-cm.

PARASITIC PROCESSES—SECONDARY BATTERIES

Besides the self-discharge processes listed above, there are a number of parasitic processes associated with recharging secondary batteries that lead to a loss in performance.

Possibly the most significant loss in available capacity is the limitation on depth of discharge. To achieve long cycle life, sealed batteries use only 10–25% of their capacity rating [A 316,317]. It has been observed that cells tend to become unbalanced when cycled in series. Some cells have higher charging efficiencies and attain full charge before other cells. If current is continued to be passed until all cells are charged, then those with the higher efficiency will be evolving gas, while the weaker cells are accepting charge. This imbalance problem is not severe in the regime of 10–25% depth of discharge.

The changes in the structure and of the true surface S of the active mass of a AgO/Zn battery in cyclic oxidation–reduction were studied by electron microscopy and double-layer capacity measurements on the discharge cycle [A 229]. S increased linearly with the charging current I_{ch} and decreased linearly as the discharging current I_d increased. Thus, increasing I_{ch} from 0.15 to 15 mA/cm² increased S by 0.3 m²/g. Increasing I_d from 0.15 to 150 mA/cm² decreased S from 0.8 to 0.5 m²/g. The addition of 1% PbO_2, or HgO to the active mass decreased S, after 2 cycles, from 2.2 to 1.9 and from 1.9 to 0.6 m²/g, respectively. Storage in a discharged state eliminated the formation of conglomerates of small crystals and formed large prisms of ZnO; the zinc surface was larger than that formed under conventional conditions. Storage in a charged state reversed the process. Increasing the temperature of storage to 35°C lowered S of the negative electrode [A 229].

Cycling zinc plates results in pronounced changes in the plate dimensions. This change can be decreased by reducing the concentration of KOH. This effect is also minimized in plates formed by Teflon-bonding the zinc particles. Apparently, there is a smaller amount of free electrolyte within the plates due to the hydrophobic character of the Teflon [A 232].

In certain secondary cells, the active material contained in the electrodes tends to cake and become less porous on cycling even though the reaction product is insoluble. As a result, the capacity of the battery is diminished. This aging effect has been noted for the cadmium electrode [A 233]. In an effort to avoid the problem at the positive, the addition of lithium hydroxide to the electrolytes of Ni/Cd, Ag/Cd, and AgO/Zn batteries has been suggested [P 401].

Dendrite Growth

One of the principal problems with cycling plates is the development of dendrites on the negative. This is usually associated with the existence of soluble discharge products. After several cycles of charge and discharge, these crystals may penetrate the dividing diaphragm or separator, making direct contact with the positive plate and short-circuiting the cell.

The physical forms of electrodeposited metals are generally a function of the rate of deposition (current), a smooth deposit being obtained at low local currents (<2 mA/in.²), dendrites at higher currents, and, finally, heavy sponge obtained at high rates. (Crystal morphology is discussed in Chapter 7.)

Dendrite Inhibition. A semipermeable membrane functions in preventing catastrophic zinc growth through reduction of the diffusion coefficient and the concentration of zincate at the growing dendrite tip. The general properties of such membranes have been discussed [A 236]. It was indicated that the migration of soluble silver was inhibited by (1) a chemical interaction with the separator and (2) a decrease in the diffusion coefficient through an increase in tortuosity. Similar thermodynamic and kinetic interactions have been observed between soluble zinc and separator materials [A 231]. For example, some zincate ion is chemically adsorbed by separators such as cellophane and PVA.

It is found that the zinc dendrites form within the separator rather than mechanically puncturing the separator. Apparently, as long as

ZnO is present in the negative compartment, the negative has enough zincate and grows where its concentration is the highest, avoiding the cellophane membrane. When the concentration of zincate decreases at the end of charge, penetration into the separator starts [A 231]. This mechanism also explains the influence of current on penetration of the membrane. At low current densities, replenishment of the external solution by diffusion through or from within the separator can occur and, thus, prevent penetration. At high currents, the solution in the vicinity of the plate is exhausted and the dendrite grows into the separator.

In order to counteract the effects of soluble ionic or colloidal particles from dispersing into the electrolyte, it has been suggested that a salting-out agent, *e.g.*, potassium ethylate or borate, be incorporated within the electrolyte. Borate (7–35%) is particularly useful and acts as a buffer for pH 9.8 [P 345].

Instead of using a separator of cellophane, the zinc electrode itself may be directly impregnated with cellulose. Cellulose xanthate is prepared according to the following reactions:

$$(C_6H_9O_4OH)_x + x\ NaOH \rightarrow (C_6H_9O_4ONa)_x + x\ H_2O$$

$$(C_6H_9O_4ONa)_x + x\ CS_2 \rightarrow (C_6H_9O_4OC\overset{\overset{\textstyle S}{\|}}{-}SNa)_x$$

This material is then used to impregnate the zinc; cellulose is regenerated with sulfuric acid, thereby precipitating a microporous material within the electrode pores [P 307]. This cellulose mechanically restricts the passage of zinc ions, thus preventing the formation of zinc trees.

Another suggestion for preventing the formation of dendrites is to add a colloid, such as gum arabic, to the plate [P 306]. The colloid migrates into regions of high current density and increases the electrical resistance of these points. This colloid migration under the influence of charging current produces a regulating action upon the current, making it more uniform throughout the area of the plate [P 306].

It has been found that a grid of copper between the zinc and silver oxide electrodes inhibits parasitic reactions and makes it possible to use larger currents on both charge and discharge. If the normal charging time for a battery without copper inserts is 30–40 hr, this time can be cut to less than 20 hr through the use of an internal copper grid and may go as low as 8–10 hr. Performance is also improved if the zinc negative is bonded to a copper sheet [P 375].

PLATE CONDITIONING

After assembly, many aqueous battery systems must be subjected to "formation cycling," an empirical conditioning procedure which is required for optimum performance. In some cases, the electrochemical processes involved in conditioning have been established.

Apparently, during the initial cycling of sealed nickel–cadmium batteries, some electrolyte is absorbed from the spacer and permanently incorporated into the nickel plate. The manufacturing problem, then, is to supply the proper quantity of electrolyte, taking this effect into account [P 325].

With silver positives, it has been beneficial to vibrate the electrolyte perpendicular to the positive during the initial electrochemical formation of the oxide structure. This has the result of dislodging the gas bubbles, thus providing a more even plate, fewer attrition losses, and faster charges [P 326]. A flowing electrolyte is also useful in removing bubbles, hastening charging [P 328].

It is possible to condition an active material by rapid charge or discharge, or both. During cycling, changes in crystal structure can take place; evidently, a fast discharge favors the formation of small crystallites. It has been noted [P 327] that an increase in performance of a AgO/Zn battery can be achieved under the following conditions: The cells are first charged at C/20 rate[3] and then discharged at the C rate; after eight cycles, the battery is fully charged. An "anti-coagulating" agent, such as LiOH, is also used.

Temperature of charging is also found to have an important effect on electrode capacity. During charging and discharging, significant temperature variations can take place, particularly in closely packed cells; for example, temperature increases of 15–25°F on charge and 10°F on discharge have been observed for some Ni/Cd systems. It has been found that the capacity of a nickel positive is temperature-dependent. Charging in the temperature range 45–55°F gives a permanent increase (in ampere-hours) of 20–30%. A less active form results if the temperature is permitted to rise above this range [P 329].

These "forming" processes are not restricted to positive plates. For example, an initial overcharge is reported [P 375] to increase the capacity of a zinc plate. Thus, for a Zn/AgO battery, where an initial

---

[3] The C rate is defined as the rate of charging (or discharging) required to obtain complete reaction in a battery in 1 hr. For example, a 12 A-hr battery charged at a current of 12 A would be receiving charge at the C rate.

capacity of 7 A-hr drops to 5.8 A-hr after 20 cycles of operation, the preliminary overcharge results in a capacity whose value at the start is 8 A-hr and is still as high as 7 A-hr after 100 cycles. This favorable result can be realized by impressing upon the electrode assembly an initial charge of approximately 150% of its rated capacity (assuming an excess of zinc over the equivalent silver mass), preferably at a slow rate, such as that corresponding to a completion of the charge in approximately 40 hr. The improvement realized is most marked if the positive electrodes are of greatly reduced thickness, preferably of the order of $\frac{1}{2}$ mm or less; excellent results have been obtained with silver plates of 0.3-mm thickness.

It has been noted that this overcharge will be without beneficial effect if carried out at a time subsequent to the first charging cycle. This may be explained by the fact that the alkaline electrolyte undergoes an appreciable modification during cycling, changing in the usual cases from an aqueous solution to potassium hydroxide to a liquid having an appreciable proportion of potassium zincate dissolved therein [P 375]. These observations are somewhat ambiguous, and it can be questioned which electrode (or possibly both) is being formed.

It has been observed that improved cycle life is obtained for Ni/Cd cells if they are subjected to a periodic conditioning cycle. Every fifth cycle is deleted and devoted instead to a complete discharge of the individual cells, i.e., a "bleed cycle." As a result, cell life is increased by a factor of 2 [A 131]. This can be due to an equalization of cells or an electronic effect, or both. A self-conditioning "memory effect" is observed in some secondary battery systems. This generally occurs after exactly repetitive cycling to no more than 40% discharge. After such cycling, the capacity of the cell decays down to the lower limit of the discharge cycle, and, if an attempt is made to discharge beyond this point, the battery polarizes badly. However, after subsequent charging the "memory" is lost and a deeper discharge is possible.

Many of the oxides, e.g., nickel oxide, silver oxide, and "peroxide," probably owe their activity to a defect structure. This memory effect observed with secondary batteries can be the result of a slow annealing of the defect structure of the unused oxide. It has been observed that the loss in performance is a loss in voltage and not a loss in capacity (A-hr), which is consistent with the defect-structure mechanism. Rapid charge would also be expected to yield a high concentration of defects, since the system is put into a highly nonequilibrium state.

The rapid discharge conditioning of silver oxide can be the result of an increase in surface area or an increase in defect concentration of the silver oxide or both.

Imposition of ultrasound during the chemical deposition of nickel oxide from solutions of $NiSO_4$ and NaOH increased the subsequent utilization of nickel. It has been shown that the formation of a precipitate in an ultrasonically agitated bath leads to a material of fine particle size. For example, in the case of forming platinum black, the ultrasonic field speeds up the rate of reduction, thus increasing the rate of precipitation and decreasing the size of the particles formed [A 114–116]. Ultrasonic waves (intensity, 0.15 or 0.3 W/cm^2) increase the current efficiency in the electrode position of copper from acid and neutral solutions and of zinc from alkaline solutions [A 117]. The physical characteristics of the plate are dependent on the intensity of the sound.

BATTERY PERFORMANCE AT LOW TEMPERATURES

The electrochemical efficiency of a battery generally decreases drastically with decreasing temperature. The low-temperature characteristic of batteries has been discussed in great detail in a recent report [A 237]; an extensive bibliography is available. The performances of a number of conventional battery systems are summarized in Fig. 5-5. The effect of temperature on half-cell potentials is modest compared to its effect on mass-transport rates. At low temperatures, the viscosity of the cell electrolyte increases and transport of reactants and of products becomes increasingly difficult; as a consequence, withdrawal of appreciable currents becomes increasingly difficult [A 23]. For example, there is a decrease in capacity of AgO/Zn batteries below −10°C due to poor charge acceptance of the silver electrode. This, in turn, is due in large part to the reduced mobility of the OH$^-$ ions in the KOH electrolyte at these lower temperatures [A 315]. If the current drain is low and the rate of mass transport is not controlling, most conventional cells perform satisfactorily even at low temperatures.

Two general approaches have been explored for improving the performance of batteries at low temperatures. One approach is to modify the electrolyte in order to increase the rates of the mass-transport process. The second is to increase the battery temperature by external heating.

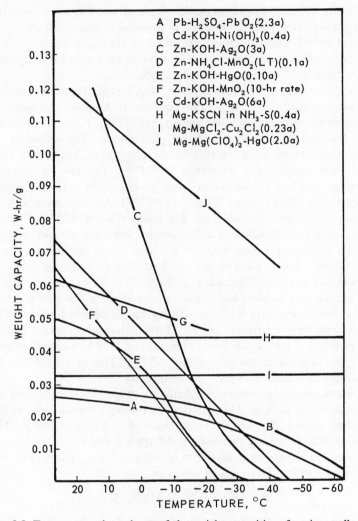

A $Pb-H_2SO_4-PbO_2(2.3a)$
B $Cd-KOH-Ni(OH)_3(0.4a)$
C $Zn-KOH-Ag_2O(3a)$
D $Zn-NH_4Cl-MnO_2(LT)(0.1a)$
E $Zn-KOH-HgO(0.10a)$
F $Zn-KOH-MnO_2(10\text{-hr rate})$
G $Cd-KOH-Ag_2O(6a)$
H $Mg-KSCN$ in $NH_3-S(0.4a)$
I $Mg-MgCl_2-Cu_2Cl_2(0.23a)$
J $Mg-Mg(ClO_4)_2-HgO(2.0a)$

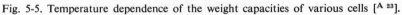

Fig. 5-5. Temperature dependence of the weight capacities of various cells [A 23].

Electrolyte Composition

A number of changes in electrolyte composition have been suggested in an effort to improve low-temperature performance. One approach is to add isopropyl alcohol (10%) to decrease the freezing point. In the case of alkaline cells, the use of RbOH or CsOH in the place of KOH improves performance of alkaline cells, since the freezing point is lowered, viscosity is decreased, and conductivity is increased [P 286]. Unfortunately, the price of these electrolytes is too high for large-scale commercial use; however, they may be acceptable for a number of low-volume military applications.

In the ordinary LeClanché cell, NH_4Cl crystallizes out at low temperatures, and $ZnCl_2$ is a poor conductor. Aqueous solutions of hydrochloride salts of simple amines (17–50%), e.g., monomethyl amine hydrochloride, give suitable electrolytes. The addition of critical amounts of NH_4Cl stabilizes the electrolyte, improving shelf life and low-temperature properties [A 70; P 241]. The low-temperature performance of the LeClanché cell can be improved with the proper ratio of $CaCl_2$, $ZnCl_2$, and NH_4Cl, [A 70; P 287]. Another method is to incorporate lithium chloride into the usual electrolyte [P 288,289]. At −40°F, ordinary LeClanché cells deliver substantially no energy, while LiCl cells were operated for 100 min. Some performance was also obtained after six-months storage. The room-temperature performance was not adversely affected.

The use of a lithium salt electrolyte to increase cell performance has also been successful with the Mg/Ag halide deferred-action cells. For example, an electrolyte of 41% LiBr, adjusted to pH 2–7 with HBr, provides a cell which has a variation of 0.14 V between 75 and −20°F for the same current density [P 290]. The corrosion of the magnesium also helps increase cell temperature.

Liquid-ammonia electrolyte batteries have been proposed as low-temperature devices. This is based on the high conductivity of the electrolyte within the temperature range of −65 to +165°F. The details of this system are discussed in Chapter 3 and 6.

Organic solvents have also been investigated, e.g., acetonitrile and propionitrile [A 238]. Dry cells of the type

$$Zn/AN, NH_4SCN/MnO_2, C$$

were built; the low-temperature limit was −50°C. At −42°C, these systems performed significantly better than the conventional dry cell (size D) [A 239]. For example, after a 60-min discharge under a constant load, the AN cell generated 6 mA, while the D cell generated 1 mA. At room temperature, the two types of cells were within 0.5 mA.

Except possibly for the liquid-ammonia battery, there are no good high-drain-rate batteries which will operate at low temperature.

Alternatives are to devise means for introducing heat into the battery and to develop effective insulation to keep the battery warm. Some techniques for accomplishing these are listed below. The aforementioned report [A 237] should be consulted for a more extensive discussion of the problem of heating and insulating batteries.

External Heating

Much of the early work on external heating of batteries was carried out with the lead–acid cell; however, the principles apply to other batteries as well.

A direct method for heating is to wrap resistance heaters around the cell and supply power from an external source. However, this is generally inefficient, since normally the battery jar is a dielectric material with excellent insulating properties.

Another approach is to imbed heaters within the battery itself; heating pads can be placed within the cell walls [P 271] or in the sludge compartment beneath the electrodes [P 272]. Better contact between the electrolyte and the heat source is achieved by the use of immersion heaters [P 273,274]. It is, of course, necessary that the heaters be clad in an electrolyte-resistant material. Coils placed in the electrolyte (1) are subject to corrosion and shock breakage, (2) take up considerable volume, and (3) require additional openings in the cover for access and connections. Metal coils are not practical for continuous immersion and plastic coils have poor heat transfer even in thin-walled tubing.

To avoid the weight of heating elements, a number of methods have been devised to heat the battery through the plates themselves. For example, an AC current supplied by an external source will increase cell temperature without consuming active material [P 275]. The AC signal can be blocked from the load circuit by the proper placement of a capacitor [P 276].

Electrical heating requires a separate power supply which may not be available under some circumstances. It is possible, in the absence of an external power source, to obtain some heating from the battery plates themselves by shorting some of the cells directly or through heating elements within the cell [P 277,278]. Spontaneous corrosion of active anodes, *e.g.*, magnesium, can also raise the battery temperature. These techniques are, of course, wasteful of battery capacity.

Another method of heating is to circulate hot air over the top and sides of the battery, which may be finned in order to promote heat transfer. The disadvantage of this method can be shown by the following calculation [P 270]. Measurements on a cell have shown that the heat transfer by such a method is about 0.005–0.01 Btu per minute per degree Fahrenheit (electrolyte to air) per cubic foot of air. The heat required to change the temperature of the cell 1°F is about 200 Btu. Hence, if a factor of 0.01 and a temperature differential of 100°F between the electrolyte and air are assumed, it would require an air flow of 2000 ft³/min to cool or heat the cell 1°F/min on open circuit. This is generally impractical.

Some of the disadvantages of electrical heating can be avoided by circulating an antifreeze solution from a heat exchanger to the cell [P 279], through the battery walls [P 280], or through the bus bars [P 270]. Water is, of course, a better heat-transfer agent than air because of its much higher heat capacity. It is appropriate to mention that these circulating systems are also useful in rejecting waste heat produced by battery inefficiencies.

A novel and effective approach to battery heating is the vest-type dry battery. This device consists of cells, made according to room-temperature formulations, arranged in thin cell blocks distributed in a canvas carrying case. The battery is worn by the user under his clothing and operates at a temperature of about +40°F. This method has established a 30-hr service level under arctic conditions, which is comparable to room-temperature capacity of conventional structures [A 24].

An alternative to electrical heating is the use of a compact chemical heating cartridge [P 281]. The heat is generated by the reaction of a readily oxidizable material, such as a metal powder, with a nongaseous oxidizing agent, such as MnO_2 or barium chromate. The heating-cartridge approach has been used extensively in low-temperature military applications. A comparison of combustion and electric heaters is given in Fig. 5-6 [A 23] in terms of system weight capacity.

Reserve batteries can be brought to optimum operating temperature by passing the activating electrolyte through a heat exchanger immediately prior to activation [P 282,283]. This type of system has application in missile batteries [A 25] where fast activation is desirable. Resistance heating is unacceptable because of a long warm-up time; often about 1 hr is required to bring a battery from −40°C to operating temperature. Some work has, therefore, been done on chemical heating. One system [A 25] has an activation time of 0.6 sec from an electrolyte and block temperature of −40°F. An example of this approach is

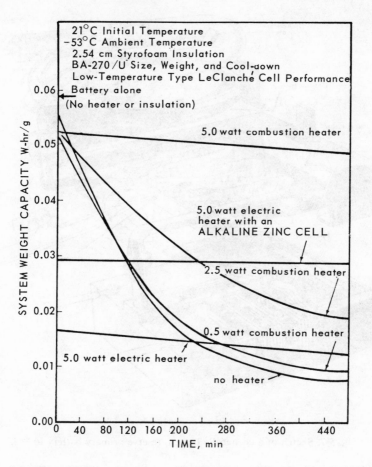

Fig. 5-6. Effect of different heaters on the performance of insulated batteries [A 23].

illustrated in Fig. 5-7 [A 290]. The effectiveness of this electrolyte–heat exchanging concept depends on the mass of electrolyte used *versus* the mass of the cell block. In one example [P 284], by heating the electrolyte to its boiling point, it was possible to raise the temperature of a battery stored at −40°C to only −15°C.

A disadvantage of this method of heating the cell is that, on initial contact of the heated electrolyte with the separators and active materials, deleterious effects result, thereby interfering with the extended performance of the battery.

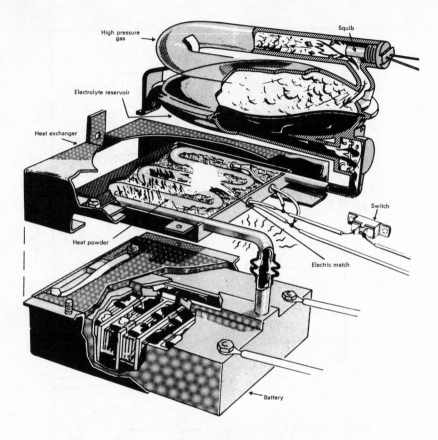

Fig. 5-7. Sketch of a chemically heated, reserve primary battery [A 290].

Internal Heating

A battery may be brought to operating temperatures through heat generated by a chemical reaction occurring within the cell compartment. For example, considerable heat is released during operation of magnesium anodes. The electrochemical irreversibility of the electrode is responsible for roughly two-thirds of the heat; the remainder comes from the exothermic corrosion. As mentioned, the theoretical E^0 for the following reaction is 2.37 V:

$$Mg \rightarrow Mg^{+2} + 2e^-$$

However, the observed potential is of the order of 1.56–1.8 V. The voltage difference appears as waste heat and is generally considered detrimental in operation at room temperature; however, this internal heating does lead to high performance at a low ambient temperature. Furthermore, the amount of heat generated increases with current drain. It has been shown that an insulated $Mg/Mg(ClO_4)_2/HgO$ battery can be operated over the ambient temperature range of $+70$ to $-40°F$ without any decrease in W-hr/lb capacity [A 240].

Attempts have been made to apply other chemical reactions within the battery itself to bring the cell up in temperature. One method for chemically heating reserve batteries is based on the heat of neutralization. An acid and a base electrolyte are stored in separate compartments. On activation, the electrolytes are mixed, heating the cell. An excess of caustic is used for electrolyte conductivity [P 282].

It has also been proposed to chemically heat an alkaline Zn/AgO battery by using an electrolyte composition which reacts exothermally with at least one of the electrodes. The electrolyte consists of KOH and an aldehyde, either alone or together with alcohol, which react with the AgO electrode [P 284]. The example quoted below is taken from the patent. "A deferred action silver/zinc battery having a rated capacity of 12.5 A-hr was stored at $-45°C$ overnight. One-hundred cubic centimeters of an electrolyte which contained 45% KOH in water having a specific gravity of 1.445 was added to the cell. This cell was used as a control. A similar cell was activated by the addition of 100 cc of an electrolyte containing 45% KOH in water having a specific gravity of 1.445 and which also contained 60 g of a 38% solution of formaldehyde. The cell temperature is set forth in Table 5-IV. An

TABLE 5-IV

An Example of Internal Heating

Time after activation (sec)	Temperature (°C)
5	−23
15	−18
30	− 6.67
60	+37.78
180	+72.20

electric load (19 A) was connected to the battery 1 min after activation. The voltage measured under load was 1.50 V. The control cell did not reach this voltage within 1 hr. The capacity yield was 70% of normal room-temperature capacity."

Insulation

The following general conclusions on insulating batteries are taken from the work of Horne and Richards in [A 23].

1. An extreme arctic condition of $-65°F$ requires the use of rather sophisticated, evacuated types of insulations, such as evacuated fibers or powders or multilayer evacuated high-efficiency insulations. Auxiliary heaters on the battery have also been used in conjunction with a more readily available foam type of insulation in order to achieve operation of batteries for more than about 5 hr.

2. The use of a modest amount (1 in.) of expanded polystyrene foam insulation can lengthen the operating life of typical communications-type battery by a few minutes to a few hours over the life expectancy of an uninsulated battery.

3. The use of external battery heaters wrapped around the battery and then insulated with a modest amount of insulating material appears to be the best way of achieving missions up to 12-hr duration.

4. Development efforts should be aimed at achieving operational types of evacuated insulation schemes for battery systems. The long mission life available with evacuated-type insulations reduces the necessity for using external heating for the battery system.

DISPOSITION OF REACTION PRODUCT

The reaction products of a plate discharge can, of course, be soluble or insoluble. In the latter case, it is necessary to provide a high-surface-area substrate to prevent blocking of the surface.

For optimum energy density at high rates, it is desirable to employ a discharge reaction with a soluble product. However, it is then necessary to provide sufficient electrolyte for direct participation in the discharge reactions and for solvation of soluble reaction products.

As an example, consider a cell discharging for 15 min at a rate of 2.5 A/in.^{2} (0.0452 equivalents/in.^{2}). It is assumed that the electrolyte will become saturated at 14 N of additional solute. The amount of electrolyte necessary to solvate the reaction product would be 3.1 cc/in.^{2} of electrode, which assumes a sufficiently high diffusion rate for rapid removal of product from the electrode surfaces to the bulk. This quantity of electrolyte per square inch represents a thickness of 5 mm. However, for minimal cell volume and minimal internal resistance loss, it is desirable that the electrode separation be of the order of 1 mm.

One answer to this problem is to provide for electrolyte circulation. This approach has apparently been used with the U.S. Navy's Mark 67 water-activated Mg/AgCl battery used in underwater applications [A 256]. It is found that the reaction products formed at the anode must be flushed out of the intercell spaces if the working voltage (1.07 V/cell) is to be maintained at high current drains ($>$1.5 A/in.^{2}). Consequently, a stream of electrolyte is forced through the battery during discharge to flush out the products and to also prevent overheating of the battery. In this particular system, the products are insoluble gels principally containing hydrated $Mg(OH)_2$ and $MgCl_2$.

Chapter 6

Weight Efficiency K_2

The weight efficiency K_2 of a battery is defined as follows:

$$K_2 = \frac{w_0}{\sum W}$$

where w_0 is the weight of the redox couples and $\sum W$ is the overall weight of the battery. $\sum W$ can be subdivided as follows:

$$\sum W = w_0 + w_e + w_s + w_a$$

where w_e is the weight of the electrolyte, w_s is the weight of the basic structural components, and w_a is the weight of the auxiliary components. This chapter will be concerned principally with the factors which constitute w_s and w_a. A discussion will be given of methods employed to construct electrodes and cells and the influence of plate size and cell size on energy density. The following specific topics are also included: (1) oversized plates or third electrodes required for sealed cells, or both; (2) the structure of reserve cells and activation systems; and (3) heat rejection systems.

PLATE STRUCTURE

The energy contained in the active materials is delivered to the terminals through the plates. The availability of this energy and the rate at which it can be delivered are a function of plate design. The theory involved in withdrawing power from a porous conductive structure was discussed in Chapter 1. The information presented below is a description of the various types of electrode and plate configurations which have been employed; an attempt is made to describe general structures rather than specific systems.

Important parameters in evaluating electrode structures are the total surface of the electrode presented to the solution, the pore size, and the tortuosity of the structure. An optimum pore structure exists which provides the minimum concentration polarization consistent with a high-surface-area electrode. Unfortunately, specific correlations of activity and structure are rarely available in the literature (*e.g.*, [A 315]).

Electrochemical Plating

Perhaps the earliest and most direct method of forming active plates is by the electrochemical deposition of active material upon a conductive substrate. The direct electroplating of metals is, of course, a well-known process and will be considered in Chapter 7. It is sufficient to note here that this method is limited to those materials which are reversible in accessible electrolytes.

Electroplating active material is not restricted to negative electrodes. For example, positive electrodes for high-rate reserve batteries have been formed by electroplating PbO_2 onto inert, conductive substrates [A 200]. Similarly, AgCl electrodes are formed by anodizing silver electrodes in chloride solution [A 201]. It is, of course, possible to form silver oxide plates from this structure by subsequently cathodizing the AgCl plate in the absence of Cl^- to form finely divided silver, followed by anodic oxidation to Ag_2O and AgO [P 396]. Apparently, it is difficult to obtain a porous deposit of sufficient thickness for use as a battery plate by electrochemical formation, particularly if the deposit is a nonconductor. It is often more convenient and effective to chemically load the plate with the proper quantity of material and, if necessary, electrochemically generate the valence state needed for discharge of the plate in a battery.

Planté-Type Plates

This configuration, originally developed for the lead–acid system, consists of a layer of finely divided, active material on a nonporous substrate. The plate is cut to develop surface and cycled electrochemically to form the porous active layer.

More recently, a similar structure has been proposed for zinc electrodes. In this case, the support plate consists of layers of expanded zinc. The cutting, shearing, and bending operations appear to roughen or pit the surface, as in etching, thereby increasing the contact area.

A limited number of cycles will develop finely divided metal [P 181]. The primary advantage of the Planté type of structure is that the electrodes are quite durable. For example, some lead batteries of this type have had useful service lives of over 30 years.

This *in situ* formation process avoids gas evolution, which occurs when electrolyte is added to a dry plate of a finely divided, active metal, *e.g.*, porous zinc, tending to disrupt the plate. The most serious disadvantage is the heavier weight, particularly with the one-piece backing plates.

Tube Electrodes

Such electrodes represent one of the early methods of forming a battery plate and are typified by the electrodes for an Edison cell. For the positive, a porous tube was filled with alternate layers of flake nickel (for conductivity) and green nickelous hydroxide. The active oxide was formed electrochemically. The positive plate swells in the forming process; the use of the black nickelic hydroxide obviates this problem. The negative plate is filled with a finely divided mixture of metallic iron, ferrous oxide, and mercuric oxide. The latter is reduced to mercury during charging and improves the plate conductivity. Tube plates for the negative electrode of the Ni/Cd cell (Junger) are filled with CdO or $Cd(OH)_2$ and are reduced electrochemically. These electrodes can be inefficient in use of material and can give rise to a high *IR* loss. Nevertheless, they do have a high net capacity and are structurally stable.

Impregnated Porous Plaques

One of the most widely used electrode structures is the impregnated porous plaque. The purpose of the plaque is to provide a mechanically stable structure which is electrically conductive and noncorrodible and which contains active material of high surface area. One common substrate is formed from carbonyl nickel and has an initial porosity of approximately 85%. The pores of the plaque can be filled from aqueous solution with the appropriate metal salt, *e.g.*, $AgNO_3$ or $Cd(NO_3)_2$. The salt is then transformed into active material by a variety of methods. For example, alkali is added to form the hydroxide, which is electro-chemically reduced to the metal. In a variation on this procedure [P 51], molten $Cd(NO_3)_2 \cdot 4H_2O$ is used for impregnation. The plaque is then reduced to cadmium metal with hydrogen. Nickel oxide electrodes can be formed similarly. A porous plaque is impregnated from a solution of 48 g nickel/100 g H_2O, dried and heated in an air–steam mixture at 180–250°C to form the proper nickel hydroxide [P 208]. The active cadmium (discharged) prepared in this manner has an area of 4.0 m^2/g, and the active nickel (discharged) has an area of 50 m^2/g [A 10].

The customary starting material in forming the plaques themselves is carbonyl nickel; the temperature, pressure, and atmosphere employed for sintering must be closely controlled to obtain a reproducible product. These plaques are expensive, primarily because of the cost of the starting material, and for this reason, some work has been done on plaques containing both nickel and iron (e.g., [P 105–107]). A nickel plate can also be formed by pressing and molding a mixture of nickel powder and 2–25% polyvinyl alcohol in hydrogen at 1500°F [P 173].

It has also been suggested [P 53] that porous carbon be coated with nickel, impregnated with active material, and then sintered. Since the density of carbon is less than that of nickel, a lighter plate should result. Another approach involves sintering the porous metal to a thin foil or screen, followed by impregnating with active material. Supposedly, these electrodes are thin, flexible, and have a high surface area [P 54–56].

The principal difficulty with these porous plates is the high weight of the inert substrate. For example, in present space-oriented Ni/Cd batteries, the support material is about 30% of the total weight of the plate [A 20]. Over 70% of the active material can be involved in cycling, but this represents only 50% of the weight of the plate.

Pressed Powder Plates

It is possible, in some cases, to form the electrode directly from porous active material by compressing a powder of high surface area. This has been described for cadmium and should apply to any cold-weldable material [P 57]. The porosity of the plate can generally be controlled by the compacting pressure [P 188]. Pressed silver plates have been used extensively. An early patent on this subject [P 58] describes a procedure whereby Ag_2O is pasted onto a nickel screen, sintered at 450°C, pressed at 0.6 ton/in.^{2} and electrochemically activated. If the screen and Ag_2O paste are heated to 500°F, only the outer surface of the particles sinter, leaving a core of high-surface-area silver oxide

[P 60]. This procedure was modified by a number of workers [P 61]. Mixing the powdered Ag_2O with water and a wetting agent apparently simplified handling the paste [P 59], although there are still some experimental problems in handling the wet paste material. Also, the sintering, with accompanying reduction of the silver oxide plate, results in certain undesirable changes in the mechanical structure and a loss in surface area.

In the dry process, the initial strength of the pressed plate (before sintering) is poor. Upon pressing, loss of surface area results, producing irregularly shaped particles and a variable, ill-defined pore structure. The problem of initial strength before sintering can be overcome by containing the active material in a temporary thermoplastic resin binder, e.g., polyethylene. The resin is subsequently burned out at 570°F [P 83]. However, there is still cracking caused by shrinkage during the sintering operation. Silver electrodes of this type also age as the metal grains slowly agglomerate. To combat this, it has been suggested that the silver be presintered [P 62] or that a mixture of silver and nickel grains be formed and sintered [P 109]. The nickel grains will "weld" together and form a reticulated skeleton which encloses the silver.

The formation of zinc plates is generally done by methods similar to those described above for other materials; for example, a grid of metallic copper, bronze, or silver [P 84] is pasted with ZnO, which is reduced electrochemically.

Expanded-Metal Grid Plates

Expanded metal has proved to be of value as a grid for pasted plates, the active material being completely or partly carried in the mesh openings of the expanded-metal sheet [P 397].

Plastic-Bonded Plates

Active metal can be permanently imbedded in a plastic [P 63], such as polyethylene [P 64]. Similarly, the active material, e.g., silver, can be mixed with a copolymer of polyvinyl chloride and polyvinyl acetate, dried, heated to 100°C, and pressed [P 65]. This procedure was modified to form a porous plaque substrate which could then be impregnated with active material. A wetting agent was used to overcome the hydrophobic character of the plastic [P 66].

Plastic-bonding active material apparently was first done in conjunction with lead plates [P 62]. For example, it was found that the electrode paste could be made by substituting 0.05–5% polyvinyl alcohol for sulfuric acid. The resulting electrode was stronger and more resistant to cracking, and the performance was as good as the conventional plates [P 93].

Another early version of this type of electrode was formed by bonding the active material, e.g., zinc, with an adhesive, such as polymethacrylate, carboxymethyl cellulose, or polyvinyl pyrrolidone [P 67,68,94]. In another example, a plastic mix is formed from polyvinyl chloride, polyvinyl acetate, and acetylene black. This is then mixed with nickel hydroxide, slurried, and applied to a metal screen. The details of the procedure are quite critical [P 110].

More recent modifications of the plastic-bonded electrodes include a leachable resin or polymer with the thermoplastic resin. The improved porosity and swelling upon leaching of the plaque reportedly improve cell performance [A 10;P 69,74]. To get high–material loadings, the procedure described above was altered to include a "metallic soap" (e.g. nickel stearate) in the mix. This improves the blending of materials and also improves subsequent electrode efficiency, presumably acting as a wetting agent. The performance of low-rate Zn/AgO batteries formed in this fashion has been described [A 9]. At the 20-hr rate, an energy density of 84 W-hr/lb was obtained for a 163.5 A-hr unit. When discharged at the 2000-hr rate, 90.8 W-hr/lb was observed.

Besides the presence of excessive plastic, this type of structure shows a number of deficiencies associated with deterioration by the expansion and contraction of the active material during cycling. These deficiencies can be substantially eliminated by using thermoplastic binders in fibrous form, e.g., copolymers of vinyl chloride and vinyl acetate [P 70]. In addition, there appear to be some manufacturing problems in keeping the low-plastic and high-density metal well mixed. One suggestion is to attach the fibers to a felted nonwoven fabric of vinyl chloride, vinyl acetate, and cellulose. These fibers project through a porous plate onto which the active material is laid down [P 175].

The use of plastics in forming battery plates has also given rise to the idea of using "metal foams" to obtain a plate of high surface area. For example, a slurry is formed of the active material plus a foam-producing and foam-stabilizing agent. The foam is then generated with air and the plate rolled to the desired thickness [P 95]. An alternate procedure is to form a porous substrate from a plastic foam. This substrate is then impregnated with active material plus wetting agent, followed by pressing and rolling to provide electrical contact [P 96]. These two patents describe application of this idea to plates for the lead–acid cell.

The plastic-bonded electrodes can be considered as either an extension or an implementation of the idea of "expanders," materials added to the active paste to prevent contraction and solidification of the spongy active mass. A number of patents deal specifically with this subject as applied to alkaline cells. For example, cellulose acetate has been used as an expander for the cadmium electrode. This material slowly hydrolyzes and swells in alkaline electrolyte, thus preventing contraction of the cadmium metal during cycling [P 71]. Polyvinyl alcohol has been used for the same purpose [P 72,73], as have rubber [P 81] and polyvinyl pyrrolidone [P 82].

Metal-Fiber Plates

It is possible to form porous plaques by using metal fibers rather than metal powders to obtain lightweight, flexible electrodes. A number of patents [P 68,75–80] describe plaques formed from fibrous copper or steel to which the active mass is bonded. Such plaques may also be formed from fibers of nickel or nickel-coated steel [P 174]. High-rate batteries can also be prepared by first spraying metal particles onto a layer of nonconductive fibrous material, such as asbestos and then impregnating the plate with active material [P 176,177,180].

Both the plastic-bonded and fiber electrodes can be spiral-wound and fitted into cylindrical cells [P 45,46]. One interesting application of flexible battery structures is their use under arctic conditions, where they may be wrapped around a person to absorb the heat necessary for optimum performance [P 36].

Printed Battery

An entirely different type of plate structure is described as the "printed battery" [P 34,35]. Thin strips of electrode (a viscous ink of graphite) are "printed" onto a porous, paper-like membrane. The depolarizer can be mixed with the ink or placed in the electrolyte, e.g., a polysulfide. Another possibility is to vapor-plate metals (e.g., zinc) onto one side of the paper.

Plastic Grids

In an attempt to reduce plate weight of heavy-duty batteries, it has also been proposed that the grid be formed of plastic and overcoated with metal. Such a structure has been described for the lead–acid cell [P 178] and the alkaline battery [P 179]. This would be most important when dealing with large plates in free electrolyte; in such an environment, extensive grid work would be needed for mechanical stability

Bipolar Plates

An important method of reducing intracell ohmic resistance is the bipolar plate. The basic concept involves the anode of cell 1 affixed to one side of a conductive plate and the cathode of cell 2 affixed to the other side of the same plate. Thus, a low-resistance series connection is made directly through the plate, without external wiring. The plate also prevents mixing of the reactants. A resin-encapsulated cell employing mechanically connected bipolar plates has been described [P 40,421]. Bipolar plates can also be formed by compression-molding of electrochemically active material through lightweight, expanded-metal grids against a thin, impervious, electronically conductive membrane [P 44].

For high-rate batteries, it is necessary to obtain minimum contact resistance between the cathode depolarizer and the cathode plate. A number of attempts at forming bipolar plates have employed spot welding to connect anode to cathode. This can be of limited succes only, since the low-resistance paths are restricted to the welds. Highly reactive oxides will often form a partly insulating oxide film on the conduction plate or grid; this has been observed for nickel and steel substrates. A good deal of the problem can be eliminated by electro-plating the substrate with a coating of silver or gold [P 184,424].

Bobbin

One of the more common plate configurations is the dry cell bobbin. Details concerning the formation of these structures can be found in the work of Vinal [A 200]. Briefly, the bobbin cylinder includes the depolarizer mixed with a conductive binder and wetted with electrolyte. This mix is then molded or compacted about a conductive carbon rod. Common depolarizers employed in this configuration are manganese dioxide, mercuric oxide, silver oxide, copper oxide, vanadium pentoxide, and nickel oxide. Materials employed as the conductive binder are carbon black, acetylene black, and graphite. Acetylene black is considered the most desirable, since it tends to form a conducting web throughout the mix and, hence, lower internal resistance [A 1]. A further decrease in resistance and improved structural stability can be obtained by including in the mix fibers of chopped or shredded steel wool, iron wool, or nickel wool [P 423].

Effect of Plate Resistance

As a result of the thin support grid and the resistivity of the active material, the net IR loss along the length of a plate will vary with position across the height of the plate. As a result, when close to the point of exhaustion of the active material, it is also possible to have different reactions taking place at different portions of the plate, *i.e.*, electrolyte decomposition and discharge of active material. [These factors have been considered on a theoretical basis for a discharge reaction within a single pore, *i.e.*, in the direction of the lines of flux (Chapter 1).]

As a consequence of this variation in current flow, there will also be a temperature variation perpendicular to the lines of current. There is inherent in this a condition for local "thermal runaway." Local heating will favor a higher rate of reaction, which in turn will mean more IR heating which will increase temperature. Eventually, however, the system will become limited by concentration polarization. It can also be shown that concentration gradients arise in a direction perpendicular to the line of current flow.

One of the major problems then is that current is being drawn unevenly from the plate materials; part of the plate is effectively overloaded, and, as a result, the battery shows excessive polarization at any given net current and an abnormally low capacity. Furthermore, this effect becomes more serious as the current drains are increased.

Effect of Plate Thickness and Cell Capacity

In choosing the appropriate electrode thickness, a balance is reached between mass-transport effects and the ratio of active material weight to total cell weight. Experiments have been described [A 9] which determined the Faradaic efficiency and energy density as a

TABLE 6-I

Plate Thickness

Plate type	Plate thickness (in.)	Efficiency (A-hr/g Ag)	Energy density (W-hr/lb)
DP	0.146	0.261	62.6
	0.092	0.289	69.8
	0.064	0.318	75.9
PP	0.146	0.291	64.1
	0.092	0.296	65.8
	0.064	0.325	70.1

function of plate thickness. In this particular example, silver oxide was the active material; pressed plates (PP) and plastic-bonded plates (DP) were used. The data are summarized in Table 6-I. Decreasing the thickness to any value substantially below 0.064 in. causes a decrease in energy density for this particular battery configuration.

The value of W-hr/lb was also a function of cell size. For example, doubling cell capacity can be achieved at less than twice the cell weight. This is shown by the data [A 9] in Table 6-II for AgO/Zn cells discharged at the 100-hr rate.

In other words, it is more difficult to build a small, high-energy-density battery than a large one.

Conclusions

A number of different types of battery plates have been discussed. However, only a small amount of direct, specific comparative data is

TABLE 6-II

Effect of Cell Size

Cell weight (g)	Cell capacity (A-hr)	Energy density (W-hr/lb)
380	38.1	71
680	79.6	82.7
1340	167.2	87.1

yet available. The impregnated porous plaque has the advantages of mechanical stability, controlled porosity, and high electrical conductivity. The major disadvantage is one of weight. The pressed plates require sintering of the particles to provide structural strength, which can lead to a low-surface-area material. Furthermore, there is a minimum thickness and loading required for structural stability. Specific disadvantages have been mentioned in the next.

The most general preparative technique appears to be the plastic-bonded plate. Structural stability should be a function of the plastic and not the plate material, and the method should, therefore, be applicable to a variety of materials. Some comparison data are available for the plastic-bonded plates and the pressed plates [A 9]. This information is given in Table 6-I and indicates a higher capacity for the plastic plates. Similar conclusions are obtained by a comparison with the impregnated plaques [A 10]. In this case, the improvement results from a decrease in electrode weight by eliminating the heavy support structure. Plastic plates can also be readily formed into thin electrodes, which may improve the reversibility of the working electrodes. Such a result would simplify charging and the construction of sealed cells. The disadvantages of this structure have been discussed.

SEALED CELLS

A necessary characteristic of many batteries is that they operate in a sealed condition. For example, batteries in space satellites are sealed to protect the system components and to obtain extended life. The experimental problem of forming a successful seal is indeed formidable; a discussion of methods used at present for constructing sealed cells is given in the work of Shair *et al.* [A 203].

The experimental problem reduces to either (1) preventing gas formation or (2) eliminating the gas formed during overcharge or overdischarge, or both. Accomplishing this with high reliability almost always involves increasing the system weight ($\sum W$), which then results in a decrease of K_2.

It is almost inevitable that, on the overcharge of aqueous electrolyte batteries, oxygen will be formed at the positive plate and hydrogen at the negative. In at least one system, *i.e.*, the positive of the Ni/Cd battery, it has been shown [A 1] that oxygen can be evolved during the charging of Ni^{+2} to Ni^{+3}. The overdischarge problem arises with multicell batteries, where often at least one cell of a stack will be weak, no matter how well the cells were matched initially. On deep discharge, this cell may reverse and effectively be driven by the remaining cells in the stack. Thus, on reversal of a Ni/Cd cell, the original negative (cadmium) will oxidize OH^- and generate oxygen while the positive (nickel) will reduce H_2O and generate hydrogen. A number of solutions have been suggested to deal with one or both of these problems.

Overcharge

Oversized Plates. Probably the earliest method for disposing of the gas evolved on charging takes advantage of the ability of molecular oxygen to oxidize the negative plate. In constructing the battery, care is taken to ensure that the capacity of the negative plate is substantially larger than that of the positive. Thus, oxygen is evolved on charging at the positive before the negative is fully reduced and can evolve hydrogen. This is a popular technique with the Ni/Cd system (the implementation of this idea is the subject of certain patents [P 1–7,43]).

The electrodes are separated by a membrane containing only enough electrolyte so that the catalyst pores are moist. This sandwich is tightly packed so that the gas evolves from the back side of the positive; the back side of the negative is exposed for access to the oxygen. If a free electrolyte is used, the negative plate extends into the gas space above the fluid [P 24]. In a battery stack, the plates are so disposed that the positive of one cell "sees" the negative of another. As would be expected, the oxygen recombination rate is significantly faster if the negative is "wet-proofed" [A 204]. Such an electrode has an electrolyte film of minimal thickness.

Internal Absorption. Another method of gas removal is based on the observation that if the separator membrane is sufficiently thin (<2 mm), the gas can diffuse through the electrolyte, *i.e.*, the solubility of gas in the electrolyte is increased. Equilibrium pressure in the cell can be as high as 20 atm [P 8,41]. In a AgO/Zn system [P 9], the electrodes were encapsulated in epoxy resin, which on hardening exerted pressure on the plates. Evidently, the oxygen produced at the positive remains in the area and is gettered by the excess silver metal. A compression pressure of 100 kg/cm² seems to be necessary; at this pressure no accumulation is noted, while at lower pressure gassing does occur

[P 48]. It is, of course, necessary that the electrodes be structurally stable at these pressures.

If both hydrogen and oxygen are evolved on charging or discharging, it is possible to include a material which will form a redox couple with the gases. An example is TiO_2/Ti_2O_3, which has been suggested for the lead–acid system [P 10]. A second example of the same general idea applied to the Ni/Cd cell is given in [P 50]. Halogen is provided in the electrolyte so that on overcharge the battery will work on the halogen cycle instead of the oxygen cycle. In the presence of Cl^-, ClO^- will be formed on overcharge and then reduced at the negative. The same is noted for Br^-. However, some oxygen also results. Fluoride ion is reportedly quite effective, although this is not to be expected from theory.

Mechanical. It has been proposed to disconnect the charging current by automatic means when it is found by exploration that the casing walls start to bulge. The effects of gas pressure on internal components have also been suggested as a method of detecting excessive gas evolution [P 13].

Third Electrode. Two other approaches to the disposal of oxygen within a cell involves the use of an auxiliary "fuel cell" electrode, electrically connected to the negative (Fig. 6-1). Oxygen produced at the positive diffuses to the auxiliary electrode, and the following overall reaction takes place:

$$O_2 + 4e^- + 2H_2O \rightarrow 4OH^-$$

The rate of this reaction is generally faster on the platinum- or silver-catalyzed surface [P 17,408] than on negatives, such as zinc or cadmium, and, as a result, higher charging rates can be used.

The negative electrode generally serves as the source of charge for the reduction of oxygen:

$$M \rightarrow M^+ + e^-$$

so that active material would be consumed, as in the direct oxidation approach [P 14–16,49,410].

It is also necessary to consider the following cell formed between the auxiliary and the positive.

$$4OH^- \rightarrow 2H_2O + O_2 + 4e^- \quad \text{(at the positive)} \quad E_3$$
$$O_2 + 4e^- + 2H_2O \rightarrow 4OH^- \quad \text{(at the auxiliary)} \quad E_2$$

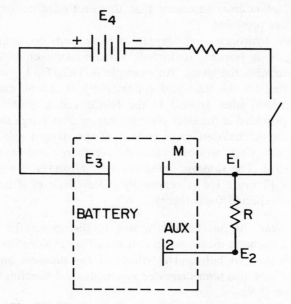

Fig. 6-1. The electrical connections for a third electrode.

The potentials at E_3 and E_2 are effectively set by the charging voltage. It is to be expected that $E_3 - E_2$ will be greater than $E_1 - E_2$. The extent that current flows from E_2 to E_3, rather than from E_2 to E_1, is determined by the relative internal resistances between the electrodes. This, in turn, is dependent on the physical placement of the electrodes within the cell. The effective difference between the two cases is that negative material is consumed in the first case, but not in the second. To some extent, this is undesirable, since it would lead eventually to hydrogen evolution at the negative.

The current developed in the auxiliary negative circuit due to the presence of oxygen can be used to activate a relay which discontinues further charging or reduces it to a low trickle rate [A 306]. This approach is based on oxygen being generated only after the battery is essentially fully charged. Successful operation has been achieved with Ni/Cd batteries. The fraction of full capacity charged into a cell using cutoff control is a minimum of 98% for charge rates from 0.5 to 3 C at temperatures of 10–42°C. The fraction of full capacity recharged decreases with increasing charging rates, declining to about 90% at the 10-C rate. In all cases, the "permanent" capacity is completely recharged. (As mentioned in Chapter 2, a small fraction of the capacity of a "Ni_2O_3" plate decays within a short time on open circuit, *e.g.*, 48 hr. This is called the "temporary" capacity.)

The use of auxiliary electrode control in batteries is more complex than for single cells. One approach is to use the signal to operate a bypass for each cell in the battery. Relays and solid-state devices can be used, but the high currents used make the bypass elements large and expensive. Nevertheless, the increase in discharge capacity can more than compensate. A less desirable approach is to use the auxiliary electrode in a pilot cell and, thus, use a single cell circuit for battery control.

In a second control mode, it is possible to use this "recombination" current to remove completely the oxygen generated on overcharge [A 307]. The electrochemical reactions are the same as shown above. The successful implementation of this approach requires a larger auxiliary electrode, all of which must be covered with a thin film of electrolyte and must be accessible to oxygen. The positioning of the auxiliary electrode may be more critical in that it is necessary that the water generated at the positive by the reaction

$$4OH^- \rightarrow 2H_2O + O_2 + 4e^-$$

be transferred to the auxiliary where it participates in the reverse, oxygen-recombination reaction.

A combination of the two approaches is probably the more effective method of operating sealed cells. It is desirable that R (Fig. 6-1) be as small as possible, consistent with developing a sufficiently large voltage for activating the cutoff relays.

Since the auxiliary is connected to the negative, it is to be expected that hydrogen will be evolved by the decomposition of water as the cell is charged. This is generally unavoidable at high charging rates. Apparently, however, small amounts of hydrogen and oxygen will safely recombine on a wetted auxiliary electrode [A 306].

Active negative electrodes, such as zinc, are relatively stable in aqueous electrolytes because of a high overpotential for the following reaction:

$$H_2O + e^- \rightarrow \tfrac{1}{2}H_2 + OH^-$$

To use an auxiliary electrode in conjunction with a negative of this type, it is necessary that the auxiliary electrode also have a high overpotential for hydrogen evolution. Auxiliary electrodes of platinum, silver, and a cobalt aluminate spinel cannot be connected directly to a zinc negative [A 205]. However, if the auxiliary electrode is connected through a diode [P 18], current will pass only in the presence of oxygen.

It has also been suggested that hydrogen be removed by connecting an auxiliary electrode to the positive. The oxidation of hydrogen is slow at these positive potentials; however, it was possible to maintain the proper voltage by the forward drop of a semiconductor diode [P 18]. A similar approach was used on the oxygen-removal electrode to avoid problems of potential-induced flooding.

Catalytic Recombination. The direct reaction of hydrogen at positives is generally slow. For example, rates of the order of 0.05 mA/in.^{2} have been observed on Ag_2O and AgO plates. The inclusion of 1% Pd increased the rate to 0.14 mA/in.^{2} in 26.5% KOH [A 313].

One of the earliest methods proposed for the removal of hydrogen and oxygen in sealed cells was catalytic recombination [A 206;P 409,411]. The approach generally involves a wire or thimble of catalyst supported in the gas space above the cells. When the catalyst fails, it generally does so because of poisoning by impurities, sintering from the heat of reaction, or flooding by an excess of reaction product. The latter problem can be overcome by supporting the catalyst on a water-repellent powder. Devices of this type have been successfully used with lead/PbO_2 cells at a controlled rate of overcharge [A 275].

Hydrogen Removal. The problem of hydrogen evolution resulting from overcharging a cadmium negative can presumably be alleviated through the inclusion of ZnO in the electrolyte. After the cadmium is charged, zinc will plate out in a thin film. A cell voltage of 1.70 V must be reached before hydrogen evolution can occur, rather than 1.55 V as in the case of cadmium alone. The plate is fully charged at 1.32 V. This scheme provides a range of 0.35 V before gassing occurs; the charging voltage should, therefore, be easier to control [P 11]. If the amount of ZnO in the electrolyte is increased to 3.5 wt.%, "trees" of zinc will plate out on overcharge and on overdischarge, shorting the cell. This effectively bypasses the cell and prevents gassing. The separator thickness is 1–15 mils. However, it is required that immediately on termination of the charge or discharge the tree be broken by oxidation of the zinc at the positive or by local action [P 12]. It is to be expected that this method will be effective only at low charging rates.

Stabistor. The development of an excessive pressure in sealed cells may be solved by holding the cell at a potential below that for oxygen evolution. This can be accomplished by shunting each cell with a "stabistor" made of an appropriate number of p–n junctions bonded together in series. The ideal device is one that has infinite impedance below the limiting charge voltage and zero dynamic impedance above this voltage. The forward characteristic of semiconductor diodes, either alone or grouped in series, approximates this characteristic [A 307].

The problems of clamping Ni/Cd and AgO/Cd cell voltages are similar due to the similarity of limiting voltages, *i.e.*, 1.48–1.52 and 1.49–1.53 V, respectively. The 1.97–2.00-V limit of the AgO/Zn couple requires a somewhat different approach.

The 1.5-V limit can be obtained by the use of two silicon p–n junctions biased in the forward direction. Because of the negative temperature coefficient of stabistor voltage, an increase in temperature of the cell at the end of charge may be used to increase the conductance of the stabistor by thermal-coupling the device to the cell. This procedure eliminates "thermal runaway" effects [A 307].

The stabistor will slowly discharge a cell on open circuit stand and will completely discharge the cell after several days. If intermittent cycling is required, then the stabistor should be removed from the cell after full charging.

Controlled Electrolyte Level. As has been mentioned, the discharge of a battery plate is favored by an excess of electrolyte. However, a minimal amount of electrolyte is desirable in recharging a battery to minimize gassing and to accelerate recombination. In practical systems, a compromise is generally reached. An investigation is now in progress to adjust the liquid level via a bellows, according to the phase of the battery cycle [A 207].

Overcharge—Conclusions. Of these methods, the use of the oversized negative has apparently been most popular. There are at least two practical drawbacks to this arrangement. First, the recombination rate may be slow, as in the case of the cadmium negative which sets the maximum safe charging rate at the 10-hr rate. Second, hydrogen may also be produced, particularly on a rapid charge; since no mechanism exists for recombining hydrogen this gas could accumulate and eventually destroy the cell.

The most obvious disadvantages of the third electrodes are the complicated construction and cost of the device. This approach would probably be most useful in batteries where the high rate of oxygen recombination (and, hence, charge rate) would be the dominating consideration. The catalysts employed to accelerate oxygen reduction have been platinum, silver, carbon, and cobalt–aluminum spinel, of which platinum is probably the most efficient.

It is to be noted that these systems can involve an explosion hazard if the electrodes become too dry. In one test case [A 206], hydrogen was reacted with partially dried AgO. The silver oxide was reduced to silver metal; the heat of reaction decomposed further AgO, generating oxygen. Finally, the hydrogen–oxygen mixture ignited on the finely divided silver.

Overdischarge

The second part of the gassing problem is overdischarge. For a maximum available energy density, it is desirable to deep-discharge a battery stack. However, it is possible that one or more members of the stack will be weak and will "reverse" during discharge. This is quite likely to occur after a secondary battery has been cycled a number of times. The result will be generation of oxygen at the original negative and hydrogen at the original positive.

Antipolar Mass. The generation of hydrogen gas with a sealed Ni/Cd cell upon polarity reversal may be avoided by employing an "antipolar" mass [P 20]. "A suitable antipolar mass is a cathodic reducible oxide, such as cadmium oxide or cadmium hydroxide, added integrally to the positive electrode. In operation, the cathodic reducible oxide is reduced at a potential lower than is required to generate hydrogen gas, and as long as a sufficient quantity of a reducible oxide is present within the positive electrode, the generation of hydrogen gas during cell reversal is prevented. In practice, the negative electrode is of such capacity and is in such a charge state that it will evolve oxygen very shortly after cell reversal occurs, thus ensuring a constant supply of oxygen to react with the antipolar mass as rapidly as it is electrochemically reduced, thereby ensuring against complete reduction of the antipolar mass." The quantity and physical disposition of the antipolar mass is the subject of certain patents [P 20–23]. Apparently, this antipolar mass is effective only at low overcharge rates [A 208].

For example, a Ag–Cd (10%) electrode will recombine oxygen (40 psia) at a rate of 10–14 mA/in.^{2} [A 209].

Electrode Sizing. It is possible to adjust the size of the electrodes so that oxygen is the gas evolved on overdischarge. This can be removed by the same general technique discussed for overcharging, assuming that oxygen will chemically oxidize the positive (or the antipolar mass).

There is a problem here, in that the relative sizes of the electrodes desirable for minimizing gassing on overdischarge are opposite that required to minimize gassing on overcharge. For example, to ensure no hydrogen evolution on overcharging, the negative electrode is made large so that it will never be charged fully and oxygen will evolve from the positive. However, to ensure that hydrogen is not evolved on overdischarge, the negative must be made smaller than the positive.

Storage. It is also desirable that gassing be avoided on wet stand. In the case of reserve batteries, this is accomplished by separation of electrodes and electrolyte until the system is ready for use. This approach is, to some extent, also applicable to secondary cells. Another answer is to store the battery in the discharged state. Application of this idea within a system assumes that (1) auxiliary power is available to charge the battery and (2) this charging can be accomplished in the time available. Tests on AgO/Zn batteries stored for 9 months in the wet discharged state showed no adverse effects except for the first cycle after stand [A 210].

RESERVE BATTERIES

The shelf life of high-energy primary batteries is often limited by self-discharge reactions between the active electrode materials and the electrolyte. One answer to this problem has been the physical separation of electrodes and electrolyte. Immediately prior to use, the battery is activated, most often by the admission of the electrolyte to the cell chamber, and operates at high power levels for short periods of time, generally minutes for fuse batteries and hours or days for signal batteries.

The physical configuration of these batteries varies greatly, depending on the application and on the forces available for activation. Many of these batteries can be activated manually. In the case of fuse batteries, activation can be accomplished by use of inertial and centrifugal forces. Often, automatic activation must be accomplished without the use of external forces; explosive squibs or thermal activation can be used. Specific examples of battery construction and cell activation will be given below. The principal design consideration for all reserve batteries is the delivery of the electrolyte to the cells at the proper time, as quickly as possible, while avoiding capacity losses through chemical short-circuiting.

General Problems

Local Action. Most reserve batteries use the bipolar plate, with anode "a" on one side of an electrically conductive sheet and cathode "b" on the other. If a film of excess electrolyte exists across the edge of the plate, local cell action develops with loss of active material. Specific answers proposed for this problem are often dependent on the cell configuration.

One way of preventing contact of the electrode edges with electrolyte is to coat the edge with an electrolyte-impermeable material. A second approach is the use of a spacer between the electrodes which absorbs electrolyte and limits the amount of fluid taken into the cells [P 231]. This leads to a slower filling time, which often cannot be tolerated in fuse batteries. In such a case, a free electrolyte must be used and some self-discharge must be tolerated.

Air Vents. A second, and sometimes more difficult, problem is disposing of the air displaced by the activating fluid. Vent holes can be provided [P 240], or the electrode compartment can be evacuated during manufacture and the cells activated by breaking a sealed port in the presence of a liquid [P 230]. However, a vacuum can be undesirable, since leakage over long storage times is unavoidable. It has also been suggested that the cells be filled with combustible gases. The battery would be activated by a heater which ignites the gases, breaking a diaphragm and expelling the gases. As the system cools, the gases contract, pulling in liquid [P 241].

Specific Electrodes and Electrolytes

Water-activated batteries generally employ seawater as the sole electrolyte and, as a result, have a weight advantage since electrolyte need not be carried along. The low conductivity of the electrolyte can be improved by the addition of solute to the dry spacer. The most commonly used electrode materials for this system are magnesium anodes [A 67] and AgCl [A 64] or CuCl cathodes. Meta-dinitrobenzene is also being considered because of its high (ampere-hour) capacity.

A wide variety of electrode materials has been employed with the electrolyte-containing reserve cells, *e.g.*, Zn/Cu [P 221]; Pb/PbO$_2$ [A 68;P 266]; Zn/AgO [A 65,66;P 251,257]; Mg/m-DNB [A 144]; Zn/PbO$_2$ [P 208,218]; Zn/SnCl$_4$ [P 202]; Zn/MnO$_2$ [P 267]; Cd/PbO$_2$ [P 216]. None of the electrode couples mentioned, except Mg/m-DNB and Zn/AgO, is of high energy density. (The specific electrochemistry of these materials was discussed in Chapter 2.)

For a rapid rate of reaction, it is desirable to employ an electrolyte in which the reaction products are soluble. In this manner, the electrode surface is continually renewed. In the lead–acid system, perchloric acid, fluoroboric acid, and fluorosilicic acid have been used, and comparative tests show these cells to have better discharge characteristics, particularly at low temperatures, than H$_2$SO$_4$ cells of the same weight and volume [A 68;P 268].

HCl is also a useful electrolyte for the Zn/PbO$_2$ couple. However,

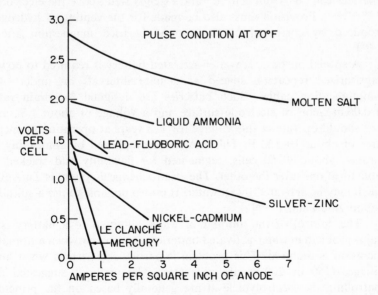

Fig. 6-2. Voltage–current density curves for batteries at high drain rates [A 143].

this battery has a short life because lead trees form on the zinc electrode as the soluble lead displaces the zinc. Apparently, this problem can be solved by the use of an electrolyte mixture, *e.g.*, HCl or HBr (0.8–9%) and 33–65% H_3PO_4. The phosphoric acid prevents the formation of lead clusters; however, it also increases electrolyte viscosity.

The electrochemical performance of a number of these couples at high drain rates is shown in Fig. 6-2.

Battery and Cell Construction

An extensive art exists on the construction and arrangement of battery elements to obtain fast and effective activation, minimum corrosion, extended life, and minimal weight and volume.

If fast activation is necessary, a free (unsupported) electrolyte should be used. A number of methods are available to maintain the proper spacing of electrodes when using a free electrolyte. Small glass spheres are imbedded within the cathode sheet [P 245] or plastic mounds can be fitted to the surface of an electrode [P 246]. Short-circuiting by the electrolyte can be minimized if a flexible, nonconductive sheet separates cell "a" from cell "b" and extends well above the electrodes [P 231,235,243]. Provision must also be made for the venting of hydrogen produced by corrosion of the commonly used magnesium anode [P 239].

A special-purpose, seawater-activated battery is required to power transitorized repeaters, spaced at definite intervals on underwater communication cables. Such batteries are designed for drain rates of 0.06 mA/cm^2 of electrode surface and a voltage of about 1 V, and they should operate at this voltage for 1–2 years at operating temperatures of about 10–120°F. This can be accomplished by connecting, in parallel, about 40–50 cells, segmented for flexibility and stacked in cable form one after the other. The electrochemical system is Zn/AgCl, which may be activated by seawater. It is also possible to use a spirally-wound zinc band [P 248].

The control of the amount of water admitted to a battery is a major problem in water-activated batteries, particularly when automatic operation is required. This problem is particularly serious when high voltages (150 or 200 V) are generated. The methods suggested for controlling the electrolyte level are generally based on the principle of inserting the battery into an envelope or hood which has an aperture.

The assembly is weighted with ballast so that, on immersion, the aperture is below the battery. After water is admitted into the hood, the hood is sealed against the flow of gases; excess water is then expelled by the pressure exerted by gas generated inside the hood. The gas may be generated artificially, for example, by contact of sodium bicarbonate and citric acid with water [P 232]. Or the gas may accumulate naturally, as the product of the electrolysis between the terminals of the battery, or by corrosion of the negative plate during operation.

The flexible rubber latex hood above the battery entraps the relatively light gases after the unit as a whole has been sunk; the battery itself is entirely submerged in the body of liquid which occupies the volume initially occupied by the gas. The excess liquid is thereafter expelled by the gases generated in the battery as referred to above. The electrolyte necessary for cell operation is retained by the capillary action of the spacer. Figure 6-3 illustrates one version of the battery immediately after immersion; Fig. 6-4 illustrates the same battery

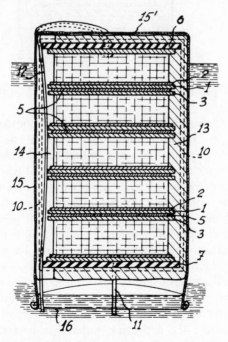

Fig. 6-3. Battery in flexible hood immediately after immersion.

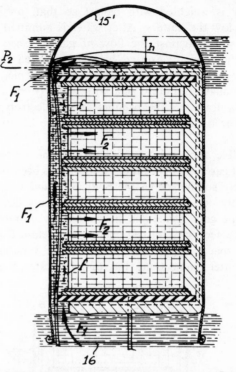

Fig. 6-4. Battery of Fig. 6-3 with hood opened [P 302].

with the hood opened. Liquid has now filled the cells and is being displaced by gas. Figure 6-5 illustrates the final state of activation [P 302]. (The numbers in these and subsequent figures, *i.e.*, Figs. 6-3 through 6-18, refer to descriptions in the original patents.)

The various types of batteries proposed differ mainly in the means used during the initial phase of total immersion for allowing the liquid to enter the hood. An electrically operated vent has been described [P 343] which will close off the battery after sufficient electrolyte has entered. Spring-activated inlet and outlet vents have also been used [P 344]. Systems involving immersion to a considerable depth are substantially complicated by the presence of a weighting mechanism [P 302].

One of the early water-activated batteries employed an aluminum anode (the outer can), a sodium hydroxide electrolyte, and a silver nitrate depolarizer; the activating water dissolved the solid NaOH and $AgNO_3$. The cell reaction was based on the oxidation of aluminum

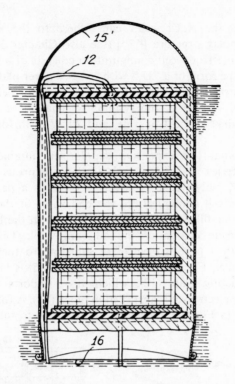

Fig. 6-5. Battery of Fig. 6-4 in final state of activation [P 302].

and the deposition of silver on the inert cathode substrate. Output was generally low [P 229]. A more popular combination is Mg/AgCl–Ag, using seawater and the reaction product $MgCl_2$ as the electrolyte [A 64]. AgBr has also been used as the positive [P 249].

Silver halide electrodes, generally thin flat sheets, can be formed by coating silver foil with silver chloride in powder form. The performance of such a cell is likely to suffer due to loosening of the silver chloride from the foil by shock and vibration. A more stable structure can be formed by using fused silver chloride [P 234,239] mechanically rolled to a wafer form. Silver chloride normally has a specific resistivity of the order of several megohms, and, even with a thickness of the order of 0.010 in., the resistance is too high. Good conductivity can be achieved by coating the wafer with a film of conductive paint; a suggested composition is 42% extremely fine silver particles, 10.8% ethyl cellulose binder, and 47.2% butyl acetate (solvent). Since the paint is conductive, the AgCl wafer can adhere to the magnesium foil, forming a bipolar structure [P 234] (see also Chapter 2).

Another electrode–cell configuration consists of a sheet of magnesium bent in the form of a "U," with a AgCl–silver plate in the middle. This plate is surrounded by porous material to absorb electrolyte [P 238]. A commercial version of this system [A 291] has an energy density of 45 W-hr/lb (dry) at the 1-hr rate and 70 W-hr/lb (dry) at the 72-hr rate.

A somewhat less expensive depolarizer suitable for seawater-activated batteries is CuCl. An electrode of cuprous chloride mixed with 10.4 wt.% polystyrene binder is applied as a paste to a bronze screen. In one cell configuration, the end wires of the screen project beyond the paste-filled area to make contact with the back face of the plate of opposite charge of the adjoining cell. This composite is assembled with a porous electrolyte substrate to form a cell in the manner shown in Fig. 6-6. The cell electrode (15) is in contact with a copper conducting plate (12), which also supports the magnesium plate (13). The electrolyte is absorbed in block (14). A four-cell assembly is also shown in Fig. 6-6. Electrolyte is admitted to the compartments

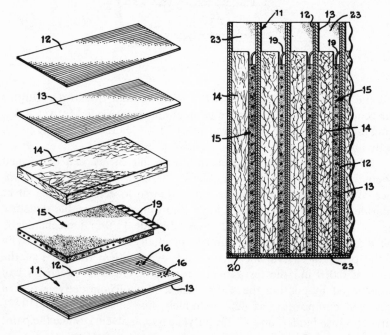

Fig. 6-6. A cuprous chloride cell construction [P 236].

(23). Note that the copper sheet (12) prevents internal shorting between the cell and the magnesium [P 236]. A three-unit structure [P 236] of 25 cells per unit with magnesium plates $0.75 \times 0.75 \times 0.014$ in. will have a wet weight of 275 g. Of this, 93 g is electrolyte, 20 g magnesium sheeting, 54 g CuCl, and the remainder is copper, fabric separator, polystyrene, and casing materials. A total discharge time of 3 hr ($2\frac{1}{2}$ hr at $-58°$F) was carried out at a load of 30 mA. A voltage of 95 V was reached within $2\frac{1}{2}$ min, and a maximum voltage of 113 V attained. The copper sheet could be dispensed with and a plastic adhesive tape, such as the copolymer of vinyl chloride and vinyl acetate, could be used on the back of the magnesium to lower the battery weight. Batteries of the Mg/CuCl type are commercially available. One set of specifications [A 291] indicates an energy density of 23 W-hr/lb of activated battery (41 W-hr/lb, dry) and a volume density of 0.86 W-hr/in.^{3} for a design life of 15 hr. After a cuprous chloride battery has been activated, it must be discharged fully within a short period of time, because the spongy cuprous chloride disintegrates in the presence of electrolyte at a relatively rapid rate (that is, over a period of several hours).

It has also been reported that a magnesium–seawater cell performs better, at low current densities, with special bacteria on its cathode than without [A 69]. Comparisons were drawn between the performance characteristics of bacterial and nonbacterial cathodes for cells otherwise identical. The biocell yielded approximately double the maximum power output. As a result, this battery system seemed a contender for long-life unattended operation at low power in an open seawater environment. However, limited life tests showed that the film-forming organisms ultimately coated the electrodes and reduced the output.

Water-Activated

Electric torpedoes and many electronic signal systems have been powered by reserve batteries (*e.g.*, [A 364]). A convenient method of activation in certain cases is the immersion of the battery in either fresh or salt water.

Automatic Activation

There are numerous applications for reserve batteries which are to be automatically activated by forces other than gravitational, inertial, or centrifugal. Batteries initiated by these specific forces will be discussed in a later section.

The liquid-release mechanism must be carried along in the system; one such device is the explosive squib. The source of power for the squib can be a charged capacitor or a long-shelf-life battery, such as the Reuben cell. Mechanisms by which the exploding squib can activate a battery will be discussed below.

Electrolyte-Release Mechanisms. The most direct use of the explosive force of the squib is illustrated in Fig. 6-7. The electrolyte compartment (18) is separated from the plate compartment (10) by a diaphragm (17). Adjacent to the diaphragm are one or more detonating squibs (23), which, when detonated, cause the diaphragm to be broken away, releasing the electrolyte into the plate compartment to activate the battery [P 250].

In another version of this approach, the squib is submerged within a completely filled container of electrolyte. The shock waves generated by the exploding squib are suddenly and fully transmitted through the incompressible electrolyte to the walls of the frangible container which rupture and allow the electrolyte to rush into the evacuated cell banks [P 251]. This electrolyte incompressibility has been used in another design. The filled electrolyte compartment is bounded by two diaphragms, one flexible and the other frangible. Pressure applied to the flexible diaphragm is transmitted directly through the fluid to the frangible diaphragm, causing rupture [P 261]. However, such a system would be particularly susceptible to temperature changes which would also alter the volume of the liquid electrolyte.

Alternatively, the squib can be used to force a knife through a diaphragm to admit electrolyte [P 250,252,253,260]. An example of such a device is shown in Fig. 6-8, where squib (34) fires knife (40) through diaphragms (44 and 58) to release the electrolyte (30) to the evacuated cell chamber (50) through orifices (58 and 41).

More rapid filling of the cells can be accomplished by a multiple-knife arrangement [P 257]. The reliability of the device should thus be increased; however, the weight of the system is also increased by the added knives.

Particularly with large batteries, the force exerted by the squib may not be adequate to fill the cells with sufficient speed. In such a situation, the squib can be used to open a valve which connects with a compressed gas cylinder [P 254,255]. This gas then moves a piston to force the electrolyte into the cells. This piston can be in the form of an inflatable bladder immersed in the electrolyte, or a sliding piston

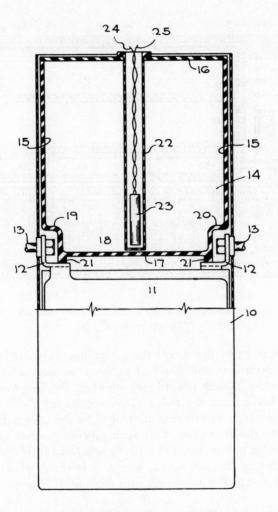

Fig. 6-7. A battery employing the explosive force of a squib as an electrolyte-release mechanism [P 250].

[P 259]. The total pressure thus applied to the electrolyte is sufficient to rupture the frangible diaphragm [P 254,263,264] (Fig. 6-9).

A novel approach to cell activation involved the following configuration. A cell was formed of a pair of parallel electrodes arranged face to face and capable of being moved toward one another. Sealing was provided around the edges of the electrodes to isolate the space

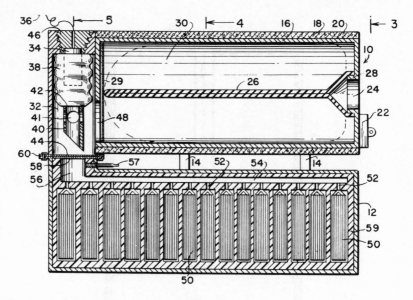

Fig. 6-8. A battery employing a squib to force a knife through a diaphragm to admit electrolyte [P 253].

between them from the surrounding medium. A frangible bag was filled with electrolyte and disposed between the electrodes. When the electrodes were moved toward one another, the bag burst and the electrolyte bridged the electrodes, activating the cell [P 262].

An alternative to bringing electrolyte to the cells is bringing the electrodes to the electrolyte. One such scheme consists of electrodes with knife-edge points stored in a cavity separated from the electrolyte by a membrane. To activate, a spring is released which forces the electrodes through the membrane into the fluid [P 265].

Gas-Activated Batteries. Liquid-activated cells have been described above. In such cell systems, the electrolyte as a whole, such as a sulfuric acid solution or seawater, is admitted to a cell already containing the other components. There are, however, certain limitations and disadvantages with such liquid-activated systems.

Many of these problems are severely aggravated by the necessity of series cell connection for most battery applications. Automatic electrolyte charging equipment, *e.g.*, frangible ampoules, in general cause temporary and, in some cases, sustained intercell shorting,

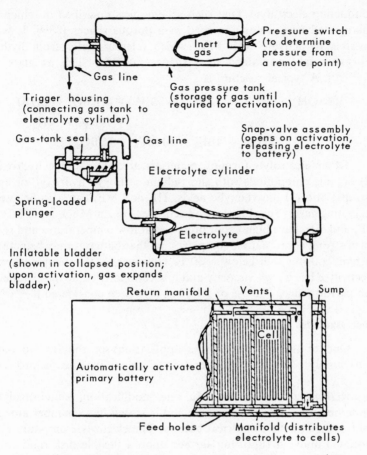

Fig. 6-9. Schematic diagram of an automatic activation system [A63].

resulting in wasted energy and noisy electrical output. Uneven or partial filling, flooding, sensitivity to position or acceleration must also be accepted unless the equipment and cells increase in complexity.

An alternative to liquid activation is the gas-depolarized cell, *e.g.*, the chlorine cathode. However, the use of this gas can result in chemical degradation of the anode. Channeling of the gas to the cathodes in series connection again causes complexity, an undesirable feature from the standpoint of reliability.

Another approach is to activate the battery by introducing into the device a gas which will react with the spacer material to form a conducting electrolyte. One such scheme was described in which HCl gas reacted with water absorbed in a porous spacer [P 218]. It is also possible to activate with BF_3, which releases water from hydrated inorganic salts contained in a spacer material, such as glass fiber [P 223,224]. A typical reaction is

$$3[Sr(OH)_2 \cdot 8H_2O] + 10BF_3 \rightarrow 3SrF_2 + 4H_3BO_4 + 6HBF_4$$
$$+ 14H_2O + 4H_2$$
$$3H_2O + 4BF_3 \rightarrow 3HBF_4 + H_3BO_3$$

Other electrolyte-forming materials are: (1) alkaline hydroxides; (2) organic compounds with short-chain carboxyl, hydroxyl, or amino groups; and (3) dicarboxylic acids. HF and SiF_4 can also be used as activating gases. Suitable electrode couples are Zn/MnO_2 and Pb/PbO_2. BF_3 and SiF_4 have high vapor pressures at low temperatures and readily form strong coordination complexes, so that batteries may be activated at temperatures well below $-30°F$. However, at this temperature, the electrolyte has a high viscosity and low conductivity, adversely affecting cell performance. The use of NH_3 has also been considered [P 225,226].

Fuse Batteries

One of the more extensive applications of the reserve battery is to supply power for the operation of proximity fuses in projectiles.

Activating Mechanisms. In one modification, an electrolyte is contained in a frangible ampoule and is stored in a chamber along the control axis of the cylindrical battery. The electrolyte ampoule can be mounted above a rupturing surface upon a flexible disk rigid enough to keep the ampoule away from the surface during normal handling, but flexible enough to allow the ampoule to collide with the rupturing surface under the influence of inertia or setback. Stresses produced in the ampoule by the impact release the electrolyte which, under centrifugal force of the spinning projectile, flows radially into the cells through filling slots provided along the inner periphery of some of the plates. Such a battery can produce a 180-V output and can have a height of 1–2 in. and a diameter of approximately 1.5 in.

Examples of such breaker devices are shown in Figs. 6-10, 6-11, and 6-12 [P 199–202]. The positions of the supporting springs and rupturing surfaces are readily apparent. Other breaker mechanisms are described in other patents [P 211–213].

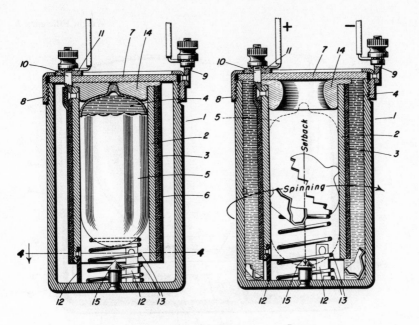

Fig. 6-10. Example of breaker device [P 202].

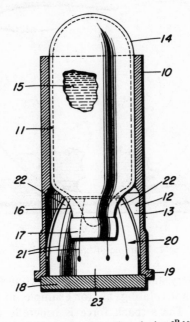

Fig. 6-11. Example of breaker device [P 199].

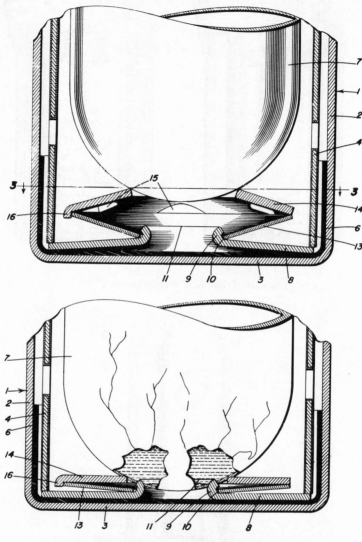

Fig. 6-12. Example of breaker device [P 201].

Filling the battery stack with electrolyte becomes more of a problem if the projectile is not rotating, as, for example, in shells fired from a smooth bore gun. The setback force is generally adequate to break the electrolyte ampoule, as discussed above. A number of proposed methods of flooding the cells make use of the capillary action of bibulous separators [P 219] or porous plates [P 220] to draw electrolyte up to the cells.

Another approach to the problem is the use of a gas-activated battery [P 218]. In one example, water was supported in a porous membrane and HCl gas was stored in the ampoule. On setback, the ampoule broke and the gas permeated the fluid, providing an ionically conducting electrolyte. The rate of activation of this battery is controlled by the rate of dissolution of the gas.

Battery Structure. The introduction of electrolyte into the cells, after rupturing the ampoule, activates the battery. Methods by which this can be accomplished are discussed in conjunction with designs taken from the patent literature.

In one general configuration [P 204,205] (Fig. 6-13), aligned grooves (14) are provided in each plate and spacer to admit electrolyte into a spacer chamber, as shown in Figs. 6-14 and 6-15. A layer of electrolyte-

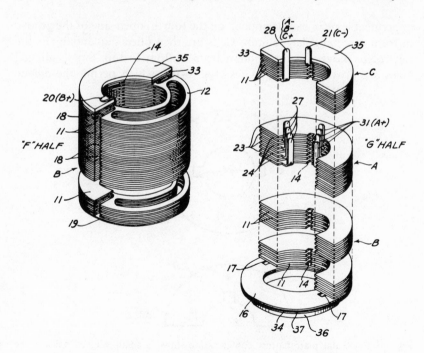

Fig. 6-13. Annular plate battery configuration [P 205].

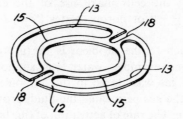

Fig. 6-14. Annular plate battery configuration showing space chamber [P 205].

Fig. 6-15. Annular plate battery configuration showing space chamber [P 205].

impermeable resin can be applied on the fourth, open side of the groove to form a rectangular column. Cells are then filled sequentially as the electrolyte moves up the column from bottom to top. Small, aligned holes are provided in the plates (Fig. 6-16), which permit the escape

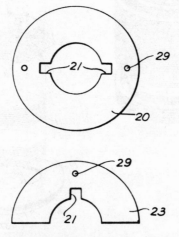

Fig. 6-16. Annular plate battery configuration showing small, aligned holes in the plates [P 204].

of air so that the cells are filled to a uniform level. Much of the successful operation of such batteries is dependent upon the correct disposition and alignment of grooves, columns and holes, and overflow vents [P 209]. Further modifications of this annular plate battery configuration are given in patents [P 214] and [P 215].

One of the principal problems with this cell is chemical short-circuiting due to local action through a film of electrolyte. For this reason, the inner portions of plates are coated after assembly with an impermeable resin. However, the column of electrolyte in the aligned holes also causes internal short-circuiting of the cells, decreasing cell life. This local action can materially reduce the watt-second capacity of the cells below their intrinsic capabilities. It appears that the flow of electrolyte through the leveling holes may often result in excessive noise voltage in the battery output. To minimize this, it is possible to choose an electrolyte so that the reaction product tends to precipitate, e.g., H_2SO_4 with lead electrodes. This is disadvantageous in that batteries using such electrolytes have a lower output.

The uniform distribution of the electrolyte to all the cells in the battery is a necessary condition to minimize battery noise. Other conditions are good mechanical rigidity of the battery and non-fluctuating electrolyte short circuits, if such short circuits are inevitable from equalizing the distribution of electrolyte.

A proposed answer [P 210] to the problem of uniform filling involves the use of an inert filler, which may be a liquid or a gas. The electrolyte from the broken ampoule displaces this filler uniformly and flow surges are constrained.

Another design of a fuse battery (Fig. 6-17 [P 207]) employs passageways (42) through which the electrolyte can flow from the spiral distribution ducts (54) to the individual cells. Upon breakage of the ampoule, the electrolyte is prevented by a barrier (44) from entering the individual cells at the "top" or inside of the battery plates. The only paths by which the electrolyte can be fed to the individual cells are through the distribution ducts provided in the container. The air enters the receiving chamber through an opening at or near the axis of rotation of the battery, so that the last cell that is filled with electrolyte will not trap substantially more air than the first. However, channel systems tend to be bulky, since the passages must be large enough to prevent clogging by dust or loose electrode material.

A third design (Fig. 6-18 [P 206]) employs involute plates. In manufacture, flat rectangular plates are initially placed vertically within an annular compartment, and the resulting "ring" of plates is then forced through a tapered sleeve to decrease gradually the outside diameter of the ring while maintaining the inner diameter constant. This causes the plates to assume an involute curve. Square separators of a suitable insulating material are positioned between the plates. A square aperture is provided in the separators, leaving only a thin wall around the perimeter of the separators and exposing a large area of the plates to electrochemical action. To prevent the separation of the plates of the final battery unit, the separators are coated or impregnated with a suitable adhesive before assembly between the plates.

Unlike a battery with radial vertical plates [P 203,208] in which the distance between plates increases with distance from the axis of the battery, the plates of this device are equidistant at all points. In a battery with radial plates, the wedge-shaped volume between plates

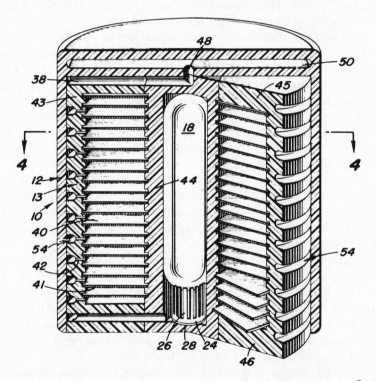

Fig. 6-17. Design of a fuse battery employing spiral distribution ducts [P 207].

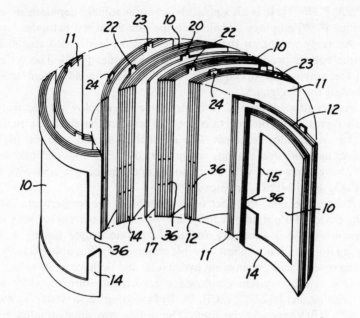

Fig. 6-18. Design of a fuse battery employing involute plates [P 206].

must be filled either with excess electrolyte or with insulating material which adds nothing to the battery.

As the projectile rotates, the frictional forces between the battery and the electrolyte cause the electrolyte to spin. As the projectile accelerates, the electrolyte also increases in angular velocity, but it attains the speed of rotation of the projectile and the battery only slowly. This differential in angular velocity is utilized to promote filling of the cells in a minimum of time. Maximum rate of fill with minimum turbulence is attained by curving the cell outward in a direction opposite to that of the projectile rotation so that the spacer ports will effectively scoop up the electrolyte.

Electrodes and Electrolytes. Most battery plates for cells of this type are of the bipolar configuration. The substrate is chosen for mechanical rigidity; steel or nickel-plated steel is quite often used. One side of the plate is coated with an oxidizable material, such as zinc [P 202,208], lead [P 215], or cadmium [P 216], generally by electroplating. The cathode side of the plate is coated with a reducible material, such as PbO_2 [P 216,218]. It is also possible to use a soluble depolarizer; one example [P 202] employs stannic chloride and a carbon cathode.

An early version of the fuse battery used a perforated zinc sleeve and the copper head of the shell as the electrodes [P 221]. The sulfuric acid electrolyte was stored in a frangible vial; an adsorbent spacer separated the electrodes.

As mentioned above, it is desirable to employ an electrolyte which forms a soluble reaction product so that the electrode surface will be continually renewed. The electrolyte should also be highly conducting; for this reason, sulfuric acid [P 208] and chromic acid [P 202] have been used with zinc anodes and fluoroboric acid with the Cd/PbO_2 and Pb/PbO_2 couples [A 60].

To some extent, the electrolyte determines the temperature range of the device. To obtain a battery that will deliver currents at very low temperatures, it is necessary to choose an electrolyte having a low freezing point and the highest possible conductivity. Possible electrolytes include several acids, potassium hydroxide, and perhaps a few salts.

One suitable system employed a eutectic mixture of water and hydrochloric acid (24.4% HCl). A hydrobromic acid–water mixture (35–40% HBr) may also be used. The anode was amalgamated zinc, the cathode PbO_2–Pb. This cell, when tested at $-55°C$, delivered peak currents of over 3 A/in.^{2} of electrode area, and it could be discharged at the rate of 0.4 A/in.^{2} for approximately 2 min, giving a terminal voltage slightly higher than 2 V at $-55°C$ [P 228].

Ammonia-Activated Batteries

The liquid-ammonia electrolyte has also been considered. The electrochemistry has been discussed in Chapter 3; cell structure is discussed below (Figs. 6-19 and 6-20).

"The FC-1 liquid-ammonia battery has four major parts: an assembly of cells, NH_3 reservoir, gas generator, and outer case. A later version, the FC-2, is similar in construction, except that the B-C section is omitted. Probably the single most difficult obstacle to packaging a complete reserve NH_3 battery was the activator section. By adopting three design principles, a solution was painstakingly achieved. The criteria were: (1) solid-propellant gas generator, (2) lightweight collapsible reservoir, and (3) mechanically pierced diaphragm. The gas generator comprises an electric squib, a pyrotechnic booster, and a solid-propellant disk. The NH_3 chamber is a thin steel shell brazed to a

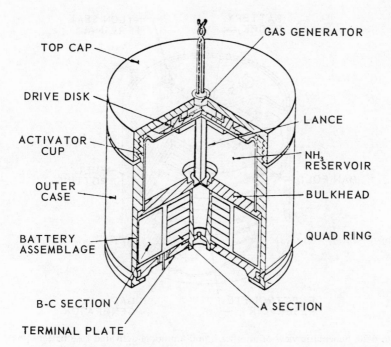

Fig. 6-19. Schematic view of a liquid-ammonia-activated fuse battery [A 144].

heavy bulkhead. Upon initiation of the squib, sufficient gas pressure is generated to collapse the reservoir, drive the lance through the bulkhead, and force passage of NH_3 into the cell stack. Residual gas pressure keeps the chamber collapsed and the NH_3 in the cells."

"The unique features of this activator design are: (1) compactness, (2) mechanical piercing of a diaphragm rather than pressure-rupturing, (3) liquid activation over the temperature range of -55 to $+75°C$ independent of the vapor pressure of NH_3, and (4) rapid functioning at all temperatures."

"The cell section of FC-2 has 15 cells in a series-connected stack that is approximately 15 mm thick and 40 mm in diameter. Each cell has a 0.1-mm bimetal sheet of magnesium/stainless steel (SST), which serves as the anode of one cell and the cathector (cathode current collector) of the adjacent cell; a 0.3-mm glass-fiber anolyte pad impregnated with KSCN; and a 0.6-mm cathode pad with m-DNB, carbon or graphite (or a mixture of both), NH_4SCN, and a glass fiber. The cathode pad may be made either by using a sheet mold or a pasting

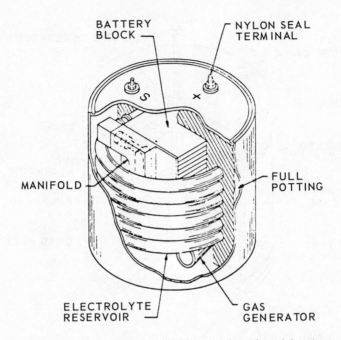

Fig. 6-20. Schematic view of another liquid-ammonia-activated fuse battery [A 98].

technique. The cell stack is sealed inside a polypropylene enclosure that has a center hole to allow free passage of the lance from the reservoir. Holes pierced in the wall of the polypropylene inner sleeve allow penetration of NH_3 into the cells. Lead wires are connected to each end cell and to hermetically sealed terminals in the battery cover plate" [A 356].

Another cell design (Fig. 6-20) adapts the power, structural, and operational characteristics of a reserve-type zinc/silver oxide battery. "The general features comprise an array of disk cells surrounded by a coil reservoir containing the ammonia electrolyte and a gas generator. This is a solution-activated design, using a neutral electrolyte, such as $KSCN-NH_3$. It should be noted that this type of coil and manifold distribution system provides a large volume of electrolyte, but requires a large quantity of gas to be generated. At reasonably high rates of discharge, the leakage current due to common electrolyte paths in the manifold is estimated to be a tolerable 10% of the total current."

"The cells for this battery are different from any others reported in that they are 8 mm thick and are a sandwich construction employing two cathodes and two anolytes around a central anode. Thus, the area of each cell is effectively twice the disk area, and cells must be wired together to form a stack. In laboratory bench tests, stacks have shown best performance in discharge for 30 to 60 min at 20 mA/cm². It is estimated that a complete battery (12 V and 1 A) could be built to yield as much as 10 W-hr/lb at that discharge time and rate. Activation would be accomplished in less than 3 sec at any temperature from -30 to $+50°C$ (-25 to $+125°F$). The complete unit, including all activator hardware, would be about 600 cm³, of which about 15% would be the cell stack" [A 356].

"Dry Tape" Batteries

This reserve battery in its simplest form consists of the apparatus shown in Fig. 6-21 [A 59]. On the right is a spool of a porous separator tape. A film on one side of this tape contains the negative material, while the positive material is coated on the other side. The electrolyte is microencapsulated and applied within either or both coatings. As the tape is fed through the crushing rolls, these microcapsules are ruptured and the electrolyte is released locally just before the tape enters the current collectors where the electrochemical reaction takes place.

Methods of electrolyte containment other than microencapsulation are possible. A series of pods could be attached to the base tape at regular intervals, which could be regarded as macroencapsulation. The periodic crushing of these pods would release the necessary

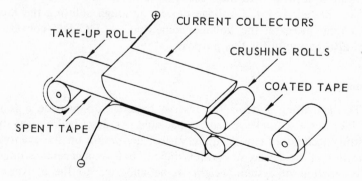

Fig. 6-21. Schematic diagram of a dry tape battery [A 59].

electrolyte. In a third method of electrolyte delivery, the electrolyte is stored separately and fed to the tape by a pumping system.

Demonstration models of the dry tape system have been built using the Zn/AgO system. The silver peroxide was bonded to a tape with polyvinyl alcohol; a second tape was saturated with electrolyte. A strip of zinc served as the anode and current collector; a gold current collector was used for the silver oxide tape. High efficiencies in the use of AgO were noted at high current densities, e.g., 92% at 100 mA/cm^2 [A 59].

Since the electrolyte and active materials are in contact for only a short period of time, it is theoretically possible to employ plate materials more unstable than could be tolerated in conventional primary and reserve batteries. For example, one system is based on the use of a picric acid depolarizer and a corrodible magnesium alloy anode. Discharge efficiencies of 50% have been achieved for the cathode at 0.5 A/in.^{2} in 2 M AlCl$_3$. The reduction efficiency is much dependent upon the wetting characteristics of the electrolyte, particularly at high loadings.

A tape couple based on the Mg/AlCl$_3$, HCl/trichlorotriazinetrione generated 2.6 V at 0.2 A/in.^{2} and 2.2 V at 0.8 A/in^2. Cathode coulombic efficiencies ranged from 60 to 80%; the discharge curve was flat and current density had little effect on cathode efficiency [A 357]. Lithium electrodes and organic, aprotic solvents are also being considered in this cell configuration.

Besides the general advantages of a reserve battery system, e.g., long shelf life, there is the added feature of start–stop operation. The entire battery need not be activated immediately, since only enough tape need be unwound to meet the specific demand.

This battery is a low-voltage device, although splitting the tape head could increase the voltage somewhat. A DC–DC converter would also be useful for this purpose.

Summary

The activating systems for reserve batteries do provide a long shelf life, but also make a heavier package. As described above, the activating techniques can be quite varied, depending on the eventual application of the device, so that it is difficult to form a general estimate of the fraction of system weight to be allocated to the activating auxiliaries. Limited information is available in the open literature on this subject. One estimate [A 60] is that "the typical (military) reserve design is probably operating at less than 25% of the energy density associated with an advanced technological form of the system as a primary battery." It has also been estimated [A 63] that the activating system accounted for 15% of the weight of a Zn/AgO missile battery.

The performance (in W-hr/lb) of a 5–12-kW battery (including the activation mechanism) as a function of discharge rate is shown in Fig. 6-22. The specific energy varies from about 18 W-hr/lb to about 32 W-hr/lb. This cell was most widely used at the inception of the ballistic missile program; overdesign had also contributed to the lower energy density.

It is to be emphasized, on quoting data from the literature, that the weight of the activating system must be included in the system weight. This is often not the case [A 61,62].

The ratio of the weight of the auxiliaries to the weight of the system improves as the power level of the battery increases. This is illustrated by the design data for the dry tape system (Table 6-III). On extended missions, long tapes can be used, so that the weight of the auxiliaries, e.g., case, drive, reels, sprockets, will be a smaller fraction

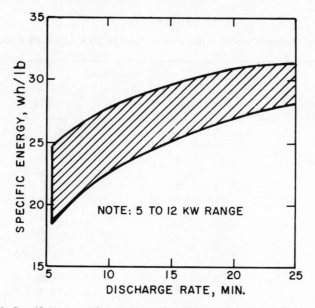

Fig. 6-22. Specific energy of remotely activated primary silver oxide/zinc batteries [A 63].

of the entire weight of the package. The considerations in Table 6-III are, of course, pertinent to most reserve batteries. Similar results are illustrated in Table 6-IV for a reserve AgO/Zn system [A 143]. In other words, it is easier to build a large, high-energy-density reserve battery than a small one.

Note also that battery capacity is a function of discharge rate. As shown in Fig. 6-23, longer discharge times lead to higher energy densities.

TABLE 6-III

K_2 (Weight Factor) for Dry Tape Battery

Mission time (hr)	Coated tape (% of total weight)
10	14
100	38
1000	49

TABLE 6-IV

Present Battery Capability Based on a Nominal 30-V Silver–Zinc Battery*

Duration	Power level			
	1 kW		10 kW	
1 min	16.7	W-hr	167	W-hr
	0.56	A-hr	5.6	A-hr
	33	A	333	A
	3.2–4	lb	17–27	lb
	5.2–4.2	W-hr/lb	9.8–6.2	W-hr/lb
1 hr	1000	W-hr	10,000	W-hr
	33	A-hr	333	A-hr
	33	A	333	A
	25–30	lb	160–220	lb
	40–35	W-hr/lb	62.5–45.5	W-hr/lb

*Taken from [A 143].

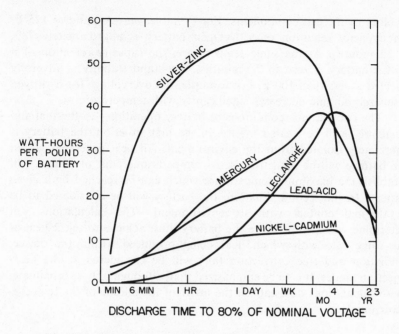

Fig. 6-23. Capacity as a function of discharge rate at 70°F [A 143].

HEAT BALANCE AND REJECTION

The heat generated by the battery is determined by the voltage efficiency and the quantity of active material discharged. The rate at which heat is produced is, of course, determined by the rate at which the battery is drained. The temperature at which the battery will operate is determined, then, by the rate at which this heat can be rejected to the environment.

The rate of heat rejection is a function of exposed surface, *i.e.*, more heat will be lost by radiation and convection from a long thin battery than from a small compact structure. Thus, it may be possible to take advantage of the fact that batteries are modular in nature, *i.e.*, the entire power plant need not necessarily be located at one place, but, for example, can be uniformly distributed throughout the walls of a vehicle.

The maximum temperature which can be tolerated within a battery module is generally dictated by the electrochemical and chemical stability of the cell components. For example, on dry stand at 125°F, the capacity retention of a AgO/Zn battery is approximately 78% in 12 months. At the same temperature, the capacity retention of a Ni/Cd battery is zero in 2.5 months. Wet-stand stability is adversely affected at less than 100°F. For example, the overvoltage for hydrogen discharge on zinc decreases significantly with temperature.

The two extreme conditions of battery operation—isothermal and adiabatic—will be treated briefly. In the first situation, the battery is completely isolated from the environment. All heat generated within the battery will be used to raise its temperature. This, of course, will establish the maximum temperature which can be reached by a given battery. In the second situation, the battery will be considered to be in thermal contact with its environment. The calculations will determine what temperature the battery must achieve to reject heat at the same rate with which it is being produced within the device. Radiation and free convection (air) will be considered as the heat-rejection modes. It is to be emphasized that, although only approximate, these calculations do establish the order of magnitude of the temperatures involved.

Adiabatic Operation

The calculation is based on the following relationship:

$$Q = W \bar{C} (T - T_0)$$

where Q is the quantity of heat absorbed in raising a battery of weight W and heat capacity $\bar{C}$ from temperature T_0 to T. The average heat capacity of LeClanché cells has been determined to be 0.31 Btu/lb-°F (Table 6-V [A 80]). A value of 0.40 will be assumed for these calculations.

Rearrangement of the above equation yields:

$$T = \frac{Q}{\bar{C}W} + T_0$$

As an example, consider a 100-W-hr battery with an energy density of 100 W-hr/lb operating at 90% efficiency. The total heat to be absorbed is 37.5 Btu, T_0 is arbitrarily chosen as 50°F, and $W = 1$. Substitution into the equation yields:

$$T = \frac{37.5}{(0.40)(1)} + 50 = 144°F$$

TABLE 6-V
Summary of Battery Specific Heat at 35°F*

Battery number	Cell type	Weight[†] (lb)	Measured specific heat (Btu/lb-°F)
BA 419/U	LeClanché	2.85	0.30
BA 2419/U	LeClanché	2.86	0.32
BA 48	LeClanché	2.77	0.32
BA 2048/U	LeClanché	4.18	0.33
BA 270/U	LeClanché	2.63	0.29
BA 270/U	Magnesium	2.09	0.37
BA 23	LeClanché	2.01	0.31
BA 35 (quarter section)	LeClanché	0.309	0.31
Mallory RM42RT-2	Mercury	0.372	0.21
			0.31 (Average)

* Taken from [A 80].
† No outside wrapper.

Note, first of all, that rate terms do not appear in these equations, so that T will be the same whether the battery is discharged quickly or slowly. Implicit is the assumption that ϵ_I remains constant, *i.e.*, the same amount of material discharges at different rates. Second, as the battery size increases, both Q and W will increase by the same factor, so that the ratio remains constant. Therefore, all other factors being equal, large and small batteries of energy density 100 W-hr/lb discharging at 90% efficiency will achieve a final temperature of 144°F.

The critical factor here is the efficiency. The same calculations, performed for other efficiencies, are summarized in Table 6-VI. In a sense, these temperatures are the result of improving the energy density of batteries, since, for a given capacity, an increase in energy density means a decrease in W. The data in the fifth column apply to a battery of energy density 10 W-hr/lb; thus, $W = 10$ lb.

These equations are also applicable to the evaluation of heat sinks for absorbing some of the waste heat. The values recorded in the fourth column of Table 6-VI represent the temperature of the high-energy-density battery in contact with an equal weight of a material

TABLE 6-VI

Adiabatic Operation of Batteries*

Voltage efficiency (%)	Q(Btu)	$T(°F)$	$T(°F)$[†]	$T(°F)$[‡]
90	37.5	144	77	59
80	85.4	262	111	71
70	146	415	180	87
50	341	850	294	130

* Values used in calculation are: $T_0 = 50°F$; $\bar{C} = 0.31$ Btu/lb-°F; $W = 1$ lb; and 100 W-hr was generated.
† Heat sink.
‡ $W = 10$ lb.

with $\bar{C} = 1$ (*e.g.*, water). The data in fifth column also apply to this same battery, but with approximately 2.6 lb of heat sink. Note that the effective energy density of the system is then 26 W-hr/lb.

Isothermal Operation

Consider a battery in thermal equilibrium with its environment. Two methods of heat rejection will be considered—radiation and convection. The temperature at which the heat loss exactly equals the rate of heat production is computed.

The Stefan–Boltzmann law describes the rate of the radiative energy rejection process:

$$q^R = e\sigma A(T^4 - T_0^4)$$

where q^R is the rate of energy loss, e is the emissivity, A is the surface area, T_0 is the absolute temperature of the surroundings, T is the absolute temperature of the body, and σ is a constant:

$$\sigma = 0.1712 \times 10^{-8} \text{ Btu/hr-ft}^2\text{-}°R^4$$

$$\sigma = 1.355 \times 10^{-2} \text{ cal/sec-cm}^2\text{-}°K^4$$

The rate loss by convection is given by

$$q^c = hA(\Delta T)$$

where h is the heat transfer coefficient for convection. For heating atmospheric air at ordinary temperatures by a horizontal square surface facing upward, h is given by [A 74]:

$$h = 0.38(\Delta T)^{1/4}$$

in which h is expressed in Btu/hr-ft²-°F and the temperature is given in degrees Rankine. Order-of-magnitude heat transfer coefficients are given in Table 6-VII [A 81].

TABLE 6-VII

Order of Magnitude of the Heat Transfer Coefficient h *

Situation	h (kcal/m²-hr-°C)	h (Btu/ft²-hr-°F)
Free convection:		
Gases	3–20	1–4
Liquids	100–600	20–120
Boiling water	1000–20,000	200–4000
Forced convection:		
Gases	10–100	2–20
Viscous fluids	50–500	10–100
Water	500–10,000	100–2000
Condensing vapors	1000–100,000	200–20,000

* Taken from H. Gröber, S. Erk, and U. Grigull, *Wärmeübertragung*, third edition, Springer (Berlin), 1955, p. 158. When h is given in kcal/m²-hr-°C, multiply by 0.204 to convert h to Btu/ft²-hr-°F [A 81].

Geometric Shape. It is next necessary to decide on a geometrical configuration, since this determines the exposed surface area which appear in the above equations. The total volume of the battery was considered fixed. Temperature comparisons will be made on an equal W-hr/lb basis; since the density is fixed, the volume must be also for a given energy capacity.

The cube was used as a reference, and only rectangular prisms were considered. The widths and heights of these configurations were expressed in terms of the length. The ratio of areas is shown in Table 6-VIII.

TABLE 6-VIII

Comparison of Surface Area for Constant Volume

Dimensions	A_x/A_0
$l = w = h$	1.00
$w = h = \frac{1}{2}l$	1.06
$w = h = \frac{1}{4}l$	1.12
$w = h = \frac{1}{3}l$	1.4
$w = h = \frac{1}{16}l$	1.67

For the sake of simplicity, in these calculations the battery will first be assumed to be in the shape of a cube. The volume, area, and weight of such a battery as a function of capacity is given in Table 6-IX. An energy–volume density of 8.5 W-hr/in.³ is also assumed.

Heat Rejection by Radiation.   We will consider first heat rejection by radiation processes alone. The emissivity is assumed to be 0.95; ambient temperature is 50°F.

$$q^R = 0.1625 \times 10^{-8}A(T^4 - 510^4)$$

$$T = \left[\frac{q^R + (510)^4(0.1625 \times 10^{-8})A}{0.1625 \times 10^{-8}A}\right]^{\frac{1}{4}}$$

Consider first a 100 W-hr/lb operating at 90% efficiency for the 100-hr rate and a capacity of 100 W-hr. The battery must reject a total of 11 W-hr at a rate of 0.375 Btu/hr. The volume of the battery is 13.3 in.³; the maximum area is 0.212 ft².

TABLE 6-IX

Battery Dimensions*

Capacity (kW-hr)	Volume (in.³)	Area (ft²)	Weight (lb)
0.1	13.3	0.212	1
1	133	0.09	10
10	1333	5.05	100
100	13,330	24	1000

* For 100 W-hr/lb and 7.5 W-hr/in³.

Thus,

$$T = \left[\frac{(0.375) + (676) \times (10^8)(0.1625) \times (10^{-8})(0.212)}{(0.1625) \times (10^{-8})(0.212)}\right]^{\frac{1}{4}}$$

$$T = \left[\frac{1.77 + (676)(0.1625)}{0.1625 \times 10^{-8}}\right]^{\frac{1}{4}}$$

$$T = [(11 + 676)(10^8)]^{\frac{1}{4}}$$

$$T = 502°R = 42°F$$

Thus, radiation processes are sufficient to prevent heating of this particular battery system. In fact, in space applications where $T_0 = 0$, T becomes $-272°F$.

This calculation was carried out for capacities of 0.1, 1, 10, and 100 kW-hr, for discharge rates ranging from 1 W to 500 kW, for discharge efficiencies of 70 and 90%, and for a sink temperature of 510°R. The data so obtained are presented in Tables 6-X and 6-XI.

TABLE 6-X
Radiation Losses (90% Efficiency Discharge)*

Capacity	Discharge rate	q^R	Temperature (°F) (vs. 50°F)
100 W-hr	1 W	0.375	50
	10 W	3.75	65
	100 W	37.5	186
	1 kW	375	585
1 kW-hr	10 W	3.75	50
	100 W	37.5	83
	1 kW	375	270
	6 kW	2250	610
10 kW-hr	100 W	37.5	56
	1 kW	375	120
	6 kW	2250	308
	10 kW	3750	383
	100 kW	37,500	1012
100 kW-hr	1 kW	375	64
	6 kW	2250	134
	10 kW	3750	177
	100 kW	37,500	551
	500 kW	187,500	1029

* Maximum adiabatic temperature, $T = 144°F$.

TABLE 6-XI

Radiation Losses (70% Efficiency Discharge)*

Capacity	Discharge rate	q^R	Temperature (°F) (vs. 50°F)
100 W-hr	1 W	1.47	54
	10 W	14.7	115
	100 W	147	383
	1 kW	1470	988
1 kW-hr	10 W	14.7	62
	100 W	147	163
	1 kW	1470	518
	6 kW	8820	1043
10 kW-hr	100 W	147	79
	1 kW	1470	247
	6 kW	8820	580
	10 kW	14,700	712
100 kW-hr	1 kW	1470	108
	10 kW	14,700	360
	60 kW	88,000	779
	100 kW	147,000	946

* Maximum adiabatic temperature, $T = 415°F$.

Of course, somewhat lower temperatures are obtained if the battery dimensions are chosen so that the area is 1.67 times that of a cube.

Isothermal Operation—Convection Losses. Heat loss by convection is given by the following relation:

$$q^c = hA(\Delta T)$$

A cubic battery attached to a device at its base will have four vertical surfaces and one horizontal surface for a total heat transfer coefficient as shown in Table 6-XII.

Consider a 100-W-hr battery rejecting 0.375 Btu/hr; the area of one face is 0.035 ft². Thus,

$$T = \frac{q}{hA} + T_0$$

$$T = \frac{0.375}{(0.035)(3.8)} + T_0$$

$$T = 53°F$$

TABLE 6-XII

Total Battery Transfer Coefficient

h (Btu/hr-ft²-°F)	T (°F)
3.8	50
4.6	100
5.44	200
6.04	300
6.54	400
6.86	500
7.26	600

These calculations were repeated for the same conditions as done for heat loss by radiation only. The results are summarized in Tables 6-XIII and 6-XIV for 90 and 70% efficiency, respectively.

TABLE 6-XIII

Convection Losses (90% Efficiency)*

Capacity	Discharge rate	q^c	Temperature (°F) (*versus* 50°F)
100 W-hr	1 W	0.375	53
	10 W	3.75	78
	100 W	37.5	247
	1 kW	375	1390
1 kW-hr	10 W	3.75	54
	100 W	37.5	86
	1 kW	375	304
	6 kW	2250	900
10 kW-hr	100 W	37.5	58
	1 kW	375	128
	6 kW	2250	360
	10 kW	3750	4840
	100 kW	37,500	3770
100 kW-hr	1 kW	375	69.2
	6 kW	2250	148
	10 kW	3750	188
	100 kW	37,500	990
	500 kW	187,500	3800

* Maximum adiabatic temperature, $T = 144$°F.

TABLE 6-XIV

Convection Losses (70% Efficiency)*

Capacity	Discharge rate	q^c	Temperature (°F) (*versus* 50°F)
100 W-hr	1 W	1.47	61
	10 W	14.7	142
	100 W	147	630
1 kW	10 W	14.7	71.5
	100 W	147	200
	1 kW	1470	1070
	6 kW	8820	5500
	10 W	14.7	60
10 kW-hr	100 W	14.7	96.5
	1 kW	1470	340
	6 kW	8820	1350
100 kW-hr	100 W	147	60
	1 kW	1470	147
	10 kW	14,700	585
	100 kW	147,000	2490

* Maximum adiabatic temperature, $T = 415°F$.

Combined Heat Rejection. Heat rejection by convection plus radiation is presented in Table 6-XV. By comparison with the data in Tables 6-X and 6-XI for radiation processes, it is seen that contribution

TABLE 6-XV

Radiation and Convection Heat Rejection*

Efficiency (%)	Capacity (kW-hr)	Rate	q_T (Btu)	Temperature (°F)
70	10	100 W	147	74
		1 kW	1470	178
		6 kW	8820	500
		10 kW	14700	620
70	1	100 W	147	122
		1 kW	1470	440
70	100	1 kW	1470	94
		10 kW	14700	300

* Maximum adiabatic temperature, $T = 415°F$.

of *free* convection does not lower the temperature significantly below that computed for radiation. At the higher temperatures, radiation processes are more effective than convection.

Summary

We have not formally considered the amount of heat rejected from the battery while it was heating up. Trial numerical solutions indicate that this correction will only amount to a few degrees for temperatures below 250°F. The numbers calculated above are of the correct order of magnitude, sufficient to justify the following conclusions:

1. Heat rejection by radiation and free convection will keep the temperature of a high-energy-density battery below the maximum adiabatic temperature for discharge rates of $\frac{1}{10}C$ or lower ($\epsilon_I \geqslant 70\%$). For net capacities less than or equal to 4 kW-hr, the maximum temperature will be $\leqslant 150°F$.

2. For high-rate discharge (and charge), a cooling system must be provided to maintain conventional batteries within the thermal stability range dictated by cell components. Forced gas flow would increase the heat transfer coefficient by a factor of 2–5. A complete evaluation of radiators and forced cooling systems is a very extensive topic and is not discussed here. To illustrate the type of result which can be expected, however, consider a 10 kW-hr battery discharged at the 1-kW rate and 70% efficiency. Increasing the heat transfer coefficient by a factor of 2 would decrease operating temperature from 340 to 195°F; an increase of 4 would provide a temperature of 120°F. Consider the same calculation for the same battery, but at the 6-kW rate (Table 6-XVI). The power to drive cooling systems must come from the battery module itself, so that such schemes are more feasible for the larger battery systems. It can be envisioned that an electrically nonconductive cooling fluid is pumped through the battery system. This could be accomplished with tubing imbedded in the battery casing, or by circulating fluid through the electrical leads. This cooling fluid is then circulated to a heat exchanger, which communicates with the outside environment. The use of fins connected to the plates for heat transfer have been suggested [P 426].

TABLE 6-XVI

Variation in Heat Transfer Coefficient

Rate (kW)	Increase factor	Temperature (°F)
6	0	1350
6	2	700
6	4	370
6	4*	210

* Also involves doubling the size of the radiator.

Space applications would require heat rejection by radiation alone. This generally means a large heat-exchanging system. This problem can become more severe for small power plants discharged at high rates, where an extensive cooling system is not feasible. The battery would have to be placed in physical contact with a high-surface-area heat sink, which would then radiate.

3. Batteries which are capable of sustaining higher temperatures can be drained at higher rates. For example, consider a 10-kW-hr battery operating at 70% efficiency in which only radiation losses are pertinent. At + 100°F, the battery will be discharging at the 200-W rate; at 240°F, the battery could be discharged at the 1-kW rate. In other words, the 10-kW-hr battery operating at 100°F is rejecting 220 Btu/hr, while the battery operated at 240°F is rejecting 1470 Btu/hr. When radiators must be provided, the hotter battery, being able to reject heat more effectively, will need smaller auxiliaries.

4. The effect of voltage efficiency should be emphasized. Consider again the 10-kW-hr battery at the 1-kW rate in which radiation losses are operative. At 90% efficiency, battery temperature is 120°F, while at 70% efficiency, battery temperature is 250°F. These differences become more pronounced at higher rates. The same battery at the 10-kW rate and 90% efficiency would operate at 400°F, but at 70% efficiency would operate at 700°F.

5. There is some advantage to adjusting battery design for a maximum of exposed area. As shown, the amount of heat rejected by radiation is directly proportional to surface area.

Chapter 7

Battery Charging

The subject of battery charging is divided here into two areas—the basic electrochemical processes involved in charge acceptance and charging circuitry. The subject of minimizing gassing on overcharge has been discussed in the section on sealed cells. Methods of preventing dendrite growth have been discussed in the section dealing with separators. We will, however, consider briefly the electrochemistry of deposit morphology, *i.e.*, the basic question of what energetic, atomic, and structural factors determine the form and properties of crystals.

The electrical forming of active plate material is essentially a problem in electroplating, and the presentation given below is taken primarily from review articles on this subject (*e.g.*, [A 326–328]). The problem considered is the electroformation of solid deposit; the treatment is applicable to both cathodic processes and anodic processes, *e.g.*, $Cd + 2OH^- \rightarrow Cd(OH)_2 + 2e^-$.

ELECTROCRYSTALLIZATION

In the electrocrystallization processes that have been examined in detail, it is observed that deposits are finely formed as discrete nuclei usually at preferred "active growth sites." Furthermore, it is observed that deposition will occur at low overpotentials. This is generally a condition of low supersaturation, so that nucleation in the conventional sense is not to be expected. The first questions then are: What is the nature of the particle being transferred, and what is the chemical nature of these growth sites?

It is generally concluded that the transferred particle is an adion and not an adatom [A 321,329,330], since the energy of activation for transfer from a hydrated ion in solution to an uncharged species is too

249

large to allow appreciable transfer rates. It is also concluded, by similar arguments, that the heat of activation in a transfer reaction in which many electrons take part is so much greater than that in which one electron takes part that the transfer of two charges in one step is improbable [A 329].

Transfer at Low Overpotentials

Now, what is the chemical nature of the growth sites involved at low overpotentials? One approach to this problem borrows from the theory of crystal growth from the vapor phase. An often used representation of this phenomenon is that atoms or molecules from the gas phase strike the surface at any point and diffuse there to a step of monatomic height, and then diffuse along this step to the "half" crystal positions or "kinks" in the step where they become incorporated into the lattice. These steps soon grow and spread over and out of the crystal surface. Once they have grown out and disappeared and a low-index surface has formed, according to the early theory, further growth will occur by means of two-dimensional nucleation of monatomic layers, which in turn would provide new steps. For two-dimensional nuclei to form at an appreciable rate, a certain degree of supersaturation is required. However, that at which most crystals grow on surfaces of low indices is much lower than required for two-dimensional nucleation. Consequently, it was suggested [A 331] that the mechanism of crystal growth at low supersaturation must involve imperfections. A screw dislocation emerging to the surface of the crystal may provide a suitable type of growth step which does not disappear during growth but is self-perpetuating.

The concept of self-perpetuating steps originating at dislocations is indispensable if low overpotentials at which metals can be electrodeposited are to be explained without invoking nucleation, which cannot yet occur at such conditions. Thus, on near-perfect crystals, such as whiskers with a low density of dislocations, ideally with one screw dislocation running along the whisker, no (copper) metal could be deposited and no growth observed, unless overpotentials were sufficiently high so that nucleation could occur [A 332].

The operation of a screw-dislocation mechanism in electrodeposition is supposedly evidenced by the formation of visible spirals and pyramids [A 333-335].

The principal differences between electrolytic metal growth and that from the vapor phase arise from the presence of the metal–solution double layer. Factors expected to cause differences are as follows [A 328]:

1. Presence of adsorbed layers, e.g., of anions or of water molecules and of metal adions,

2. Existence of the electric field with strengths of the order of 10^7 V/cm between electrode and ions in the double layer.

3. Charged nature of the particles arriving at the surface.

4. Nature of the metal particles on the surface prior to incorporation into the lattice (atoms and adions for vapor and electrolytic growth, respectively) and their interaction with the substrate.

5. Rate of the arrival of ions to the substrate. Diffusion in solution is slower and, hence, will have a greater tendency to be the rate-controlling factor, rather than the rate of arrival of atoms from the vapor.

More specifically, the *energy* of metal surfaces in contact with an electrolytic solution differs from that in contact with its vapor. Particles adsorbed from the solution reduce the surface energy. Owing to ionic hydration, the equilibrium concentration of adions of a metal in contact with an electrolyte should be greater than the adatom concentration for the same metal in contact with its vapor (not at high temperatures). The mobility of adions will be hindered by the presence of other adsorbed species on the surface, particularly by water molecules. The hydration of adions weakens their bonding to the substrate and decreases the energy of activation of surface diffusion from one site to the other. Conversely, the presence of adsorbed water molecules will make the entropy of activation more negative. The adsorption of species on the surface and the characteristics of the double layer will be influenced by the composition of the solution and by the current passing through the double layer, and these may directly or indirectly, through change of adion concentration, mobility, *etc.*, affect the mechanism of electrodeposition and the type of electrolytic crystal growth.

It is argued [A 329] that the actual transfer of an adion from solution to the substrate occurs at metal *planes* and that this adion then diffuses across the surface to the growth site. Transfer at kink sites and defects (*i.e.*, directly to the growth sites) on the metal could be quite difficult, because of the energy needed to distort the ionic hydration shell when the ion attempts transfer in a position in which the final state involves high coordination by metal lattice ions (*e.g.*, at a kink site). However, it has been difficult, experimentally, to distinguish between the surface-diffusion mechanism and direct deposition at the growth site [A 336].

As discussed in Chapter 1, an increase of overpotential decreases the *effective* activation energy for a charge-transfer process. Within the terminology employed here, an increase in overpotential should increase the number of screw dislocations reaching the surface or "activate" a number of dislocations, or both. The application of this concept to the rate equations is given in the work of Fleischman and Thirsk [A 336].

Nucleation

After application of a constant overpotential to an ideally flat surface, net ion flow across the double layer takes place, but only until the adion concentration has reached a certain value. Except for the effects of removal of adions by diffusion to, and incorporation in, growth steps, no current will flow unless the adion concentration corresponding to the applied potential is such that *nucleation* becomes possible. For example, numerical calculations show that significant nucleation may be expected in the case of silver deposition if the overpotential is of the order of 80–100 mV. Experimentally, it is found that, with copper whiskers which are supposed to be bounded by ideally flat surfaces with no steps and growth sites on them, there is no deposition current unless overpotentials of the order of 100 mV are reached. The current density which would be observed just after the commencement of nucleation is governed by surface diffusion or transfer control and can reach significant values, as the nucleation has provided necessary growth steps to which ions can diffuse.

The probability of nucleation on a stepped surface is much less than that on the atomic flat surface, as the adion concentration, even at the midpoint between steps, is far less than that at a step-free surface under the same conditions. Analysis shows that, on a surface with a high density of dislocation ($N \cong 10^{10}$ cm^{-2}), no nucleation is possible unless cathodic overpotentials are over 150 mV.

Basic Equations. The cathodic and anodic growth of electrode deposits can occur in a variety of ways [A 337]. Two extremes of behavior can be differentiated. Deposition may take place at discrete centers, the centers having a definite geometry of growth, or deposition may take place at a "uniform" surface, this surface growing into the solution. Examples of the first type are the initial stages of the deposition of an insoluble salt of a metal on an electrode of the same metal, and the initial stages of deposition of a conducting phase onto an inert substrate. Examples of the second type are the growth of a uniform metal deposit, or of a continuous oxide layer by high-field conductivity. The dividing line is not sharp and the outward growth of a nonpassivating layer could be cited as an example of uniform growth at discrete centers.

A problem in forming the rate equations for the electrodeposition is to account for the active surface area. As mentioned, not all the surface of an electrode is active. Secondly, as the nuclei grow, the active surface area increases in accordance with the geometrical configuration of the nuclei. The usual approach to this problem is discussed below.

It is necessary to divide the treatment into the initial growth stage and the steady state. In the initial stage, growth centers are formed on the electrode; this is often followed by a one-dimensional buildup of the deposit out from the surface. The formation of new centers is controlled by the *rate of nucleation* and the geometry of growth determines the *rate of increase of surface area*. Both these rates are dependent on the overpotential.

The growth of a single center may follow a variety of laws—one- and two-dimensional, if deposition is confined to certain crystallographic directions, and three-dimensional if all surfaces grow uniformly.

For the current i at time t in a single center at constant potential, we obtain, therefore, in these cases, the equations

$$i = B_1 \tag{7-1a}$$

$$i = B_2 t \tag{7-1b}$$

$$i = B_3 t^2 \tag{7-1c}$$

where the constants B are potential-dependent. If the shapes in the second and third cases approximate to cylinders and hemispheres,

$$B_1 = kS \tag{7-2a}$$

$$B_2 = 2\pi l k^2 \left(\frac{M}{FD}\right) \tag{7-2b}$$

$$B_3 = 2\pi k^3 \left(\frac{M}{2FD}\right)^2 \tag{7-2c}$$

where k is the rate of deposition on unit surface in $\text{Å}/\text{cm}^2$; S is the area of the growth [equation (7-2a)]; l is the height of the centers [equation (7-2b)]; M is the molecular weight; F is the Faraday constant; and D is the density. More complicated cases arise if deposition on a face is followed by migration to another.

The number of centers also varies with the time, being controlled by the rate of nucleus formation. In the simplest cases, this rate will be determined by the surface available for further deposition, or more usually the number of preferred sites on which nuclei have not been formed. In these cases, the number of nuclei in the initial stages will be given by

$$N = A't \tag{7-3}$$

where A' is a nucleation constant. On the other hand, if nucleus formation at preferred sites is very fast, the only law that could be observed would be the instantaneous formation of N_0 nuclei. The overall rate of reaction is obtained by combining equations of a type similar to equations (7-2) and (7-3). For these, one obtains the following:

$$i = A'B_1 t \tag{7-4a}$$

$$i = \frac{A'B_2 t^2}{2} \tag{7-4b}$$

$$i = \frac{A'B_3 t^3}{3} \tag{7-4c}$$

The cubic law [equation (7-4c)] is followed in the oxidation of lead sulfate to lead dioxide, and the interpretation is free from ambiguity. On the other hand, if nucleation is rapid,

$$i = N_0 B_1 \tag{7-5a}$$

$$i = N_0 B_2 t \tag{7-5b}$$

$$i = N_0 B_3 t^2 \tag{7-5c}$$

The time dependence in the initial stages of growth will, therefore, be usually controlled by the product of two constants, each of which may vary with potential.

It is found that the nucleation constants vary very rapidly with the overpotential, being determined by an equation of the form

$$A = Ki_0 \exp\left(\frac{-k\sigma^3}{\eta^2}\right) \tag{7-6}$$

where K is a frequency factor; i_0 is the exchange current; k is a constant determined by the shape of the nucleus, and σ is the surface energy. The variation of the constant B with potential, on the other hand, would be expected to be logarithmic:

$$B = Ck^n = C\left[k_0\pi_j a_j^{v_j}\exp\left(\frac{\alpha\phi F}{RT}\right)\right]^n \tag{7-7}$$

where C is a constant depending on whether $n = 1, 2,$ or 3; k_0 is the specific reaction rate constant; and $a_j^{v_j}$ determines the stoichiometry of the slow step of the reaction.

The experimental separation of the rate constants can be accomplished sometimes by maintaining the overpotential for a short, determined time at a high overpotential, so as to "preform" a large number of nuclei, and then to follow the growth of these nuclei at a second, lower overpotential. The nuclei formed at the higher overpotential will not have grown appreciably during the time the potential has been held at this value and all may be assumed to be of the same age. In comparison, further nucleus formation at the lower overpotential may be neglected, and the rate should follow equations (7-5a), (7-5b), or (7-5c) with N_0 replaced by N', the number of nuclei formed at the upper overpotential. The power of the time in these expressions, therefore, shows directly in how many dimensions the centers grow. Furthermore, since the constant is proportional to N', the variation of N' with the time the electrode is held at the upper overpotential can be examined.

Examples of this approach are given elsewhere [A 337]. At steady state, the equations become time-independent [equation (7-5a)]; growth is essentially restricted to a direction out from the surface.

The potential dependence of the nucleation rate constants has been tested using the values determined kinetically [A 337] or by counting the number of nuclei formed in a short interval of time at a constant potential by means of electron micrographs. The form of equation (7-4) has been confirmed cathodically for silver, lead, and mercury and anodically for the formation of α- and β-lead dioxide under a number of different conditions and for the electrodeposition of manganese dioxide. It is found, in general, that only a small proportion of the surface is available for nucleus formation. The same surface may also allow the formation of a different number of nuclei for different processes, for example, 10^9 for lead dioxide and up to 10^{12} for manganese dioxide. Different sites may be involved. It has also been found that, in the deposition of mercury and silver on platinum single-crystal spheres, the octahedral planes are preferred to the cubic planes. In the case of silver, this does not, however, correspond to any change in the surface energy term. For the electrodeposition of α- and β-lead dioxide, the effect of electrode potential and of overpotential has been examined over a wide range. It has been found that a single frequency factor fits both processes in a number of solutions, but that this factor is 10^5 times larger than the theoretical value. The values of the composite surface energy terms deduced by applying equation (7-6) are reasonable and increase with electrode potential in the manner expected. In some cases, these surface energies indicate a retardation of nucleation due to adsorption of ions on the substrate, i.e., an inhibition.

Overlap

It is observed that, after the initiation of nucleation, the rate of deposition increases. This is due to the increase in the number of nuclei on the surface. It is apparent that this current–time variation cannot continue indefinitely and still follow the ideal cases derived above, since the growth centers cannot continue freely in all directions and will impinge on each other or else on the surface of the crystallites of the parent phase. For the formation of solid phases, growth must stop at the points of contact. This situation also occurs in anodic crystallization. In the case of electrodeposition, this limitation on the size of the centers is confined to the axes parallel to the plane of the substrate, and the current must approach the steady-state value corresponding to outward growth perpendicular to the surface. In the case of oxidation or reduction of a solid parent phase or in the formation of salts on electrodes, the size of the centers becomes limited in all directions and the current can go through a maximum and either asymptotically approach zero or reach a steady state characteristic of continued outward growth. In order to calculate the current–time transient in this region, it is necessary to estimate the extent of overlap for the model chosen.

This has been done for a number of models, e.g., cylindrical, spherical, and square centers. An example in which overlap has been taken into account is the cylindrical growth of N_0 centers of argentic oxide instantaneously on silver sulfate. The centers appear to be distributed over an appreciable volume, *i.e.*, they are not confined to a narrow plane parallel to the metal substrate. However, the growth is restricted to the original sulfate crystal. An extreme example of centers interacting with the bounding surfaces of the parent phase is observed in the growth of β-lead dioxide in small lead sulfate crystals, where on the average only one nucleus is found to grow per crystal of the parent phase.

Summary

Thus, the following sequence of steps has been determined: At the initial stage of charging (low η), material deposits at (or near) growth sites. The current may then be a function of the square of overpotential. At higher η, the concentration of ions is at supersaturation and nucleation can take place. Thus, the number of sites increases with time and, hence, the current increases with time, *e.g.*, according to the equations given.

At later times, the nuclei may interfere with each other and the rate can drop or level off. At this point in time, the current must approach the steady-state value corresponding to outward growth perpendicular to the surface. It is to this steady state that i–η equations (given in Chapter 1) apply.

Examples of the experimental and mathematical analysis of the data obtained during this nuclei growth period is discussed elsewhere [A 337–339].

MORPHOLOGY

The surface energy term is of course related to the lattice binding forces (see below). It is possible to consider the work (or free energy) involved in forming a two-dimensional nucleus in terms of the lattice energies of the various crystal planes making up the substrate and the deposit. A general equation of the following form results [A 341]:

$$W_{hkl} = \frac{B_{hkl}}{(1/mN)(\mu - \mu_0) - A_{hkl}} \tag{7-8}$$

where W_{hkl} is the work of formation of a two-dimensional nucleus along the *hkl* plane; B_{hkl} and A_{hkl} are functions of the lattice energies, taking into account the interactions between first, second, and third neighbors; and $\mu - \mu_0$ is the change in chemical potential. It is this latter term that is influenced by potential and provides the rationale for the observed variations in crystal habit with changes in potential. Experimentally, it is well known that changes in orientation are inseparable from changes in the rate with which the deposits are laid down, which are determined by the overpotential.

Dendrites

"A dendrite is a skeleton of a monocrystal, consisting of a stalk and branches, and occupying two or three dimensions. The angles between the stem and the branches, or between different branches, assume only certain definite and usually simple values in accordance with the space lattice. Quite often the stem and branches represent just one lattice direction. The dendrite directions are the closest-packed lattice directions; the shape of the electrodeposited dendrites is mainly determined by the directions of preferred growth in the lattice" [A 340].

It is generally considered that the principal factors controlling dendrite growth are: (1) overpotential; (2) mass transport of soluble reactant; and (3) inhibitors.

As shown above, the overpotential affects or determines the preferred direction of crystal growth. As the overpotential is increased, preferred growth of one crystal direction decreases and a random distribution of crystals is to be expected [A 341]. It is to be emphasized that the term overpotential, as used here and in the considerations above, refers solely to *activation polarization* and does not include mass-transport effects from the bulk of solution to the double layer, *i.e.*, concentration polarization.

Mass-transport effects become important after the concentration of ions in the vicinity of the electrode becomes depleted. Ions diffusing in from the bulk of solution will now reach the dendrite extending out from the electrode surface before reaching the substrate. As a result, deposition at the dendrite tip will be favored.

The third factor involved in dendrite growth, in addition to potential and ion concentration, is the presence or absence of inhibitors. As the name implies, these are materials which compete with the adion for active growth sites. Since the growing tip of a dendrite is an active site, it is to be expected that the inhibitors will adsorb at that position, retarding further growth. The basic principles of inhibition of electrocrystallization are given elsewhere [A 342].

"These effects are illustrated in the electrodeposition of lead from 0.5 M lead acetate and 2M acetic acid. At low current densities (hence, low overpotential), the dendrite formed branches along the cube face diagonal. At higher current densities, the dendrite shape changes to rows of truncated octahedra, i.e., a dendrite branching along the cube edges in a cubic lattice. With increasing overvoltage, the tendency toward formation of dendrites is greatly retarded; likewise, coarser forms of deposit are developed" [A 340]. The dendrites formed in nitrate and acetate baths are small and delicate and without metallic luster. They appear as a black sponge and are an example of so-called "metal blacks." In electrolyte containing 2M ammonium acetate, a black spongy or even slimy deposit was obtained at 0.1 $\text{Å}/cm^2$ and consisted of very small dendrites with tertiary branches which were only about 1μ or less in diameter.

Since all the branches are perpendicular to each other, the dendrites are crystalline, not amorphous, as is sometimes stated. Formation of a black sponge was also very pronounced in an alkaline electrolyte. These anions, nitrate and acetate, are considered to be inhibiting. Gelatin also retards the formation of dendrites somewhat, although not to the same extent as strongly inhibiting anions [A 340].

The following four types of deposit observed with zinc in potassium hydroxide are listed according to increasing current density of formation [A 230]:

1. Smooth deposits—dense, with no pore structure visible at a magnification of 1000 ×.
2. Foam (light sponge)—soft deposit of low density and of very fine porosity.
3. Dendrites—hexagonal, elongated crystals with visible branches.
4. Heavy sponge—porous, dense, and hard.

The smooth deposits are obtained at low local currents (<2 mA/in.^{2}); the dendrites are obtained at higher currents; and, finally, heavy sponge is obtained at high rates.

BATTERY CHARGING SYSTEMS

The design of any given battery charging system is, of course, governed by the charging method to be employed; the charging method is governed by the electrochemical characteristics of the particular battery cell to be recharged. The general discussion in this section will thus be confined to some characteristics of known battery systems and to methods used for their charging.

Charging Constraints

For any given battery couple, the charging system must operate subject to certain constraints imposed by the physical and chemical properties of that couple. Some examples of constraints are as follows:

1. The imperfect mechanical formation and, hence, buckling of plates when charging rates are too high or battery temperatures are allowed to rise excessively.

2. A reluctance to accept fast charging due to the formation, in the charging process, of a high resistance, semi-insulating layer on the active battery plates.

3. Chemical changes in the battery electrolyte which change the battery internal resistance during charging and lead to possible conditions of negative resistance.

4. The evolution of gas when charging is nearing completion, requiring a reduction in charging rate in order not to damage the battery.

At the present state of the art, this evolution of gas is a particularly serious problem. In sealed cells, the gas can cause rupture of cell cases. In vented cell applications, where cells are enclosed within a closed system, gas evolution can cause rupture of the container or bring about an explosive atmosphere of hydrogen and oxygen mixtures within the container. In the latter case, ignition sources, such as relays or high-voltage arcing, can jeopardize the mission and be hazardous to personnel.

Control Modes

Battery charging methods generally may be classified as open-loop or closed-loop systems. Closed-loop systems sense some characteristic of the cell which indicates the completion of charge (*e.g.*, gas evolution, cell temperature rise, cell terminal voltage, and specific gravity of cell electrolyte) and use this information to terminate charging or change charging mode.

Open Loop. Open-loop systems operate for fixed time periods at fixed charging modes and terminate on the basis of charging time without having actually established whether indeed the battery has been charged or overcharged.

Closed Loop. Closed-loop systems are usually more complex than open-loop systems, but are to be preferred as a surer method of establishing full charge (providing the battery gives some indication of when it is fully charged). Closed-loop charging systems have been built using the first three characteristics listed above for nickel–cadmium cells.

Oxygen Sensing. A third electrode was added to the nickel–cadmium cells which senses the oxygen evolution near completion of charge and uses the voltage so developed either to dump further charging input outside the cell in a load resistor, or else to generate an electrical control signal which terminates charging. Such third-electrode control is quite sensitive to the battery temperature, however, causing a reduction to only 20% of full charge for a change in temperature from 40 to 140°F.

Temperature Sensing. Battery temperature can also be sensed, and the sudden rise in cell temperature near end of charge is used to terminate the charging. This method was found to be effective in preventing overcharging of nickel–cadmium batteries designed for satellite charge–recharge cycling.

Voltage Sensing. In still another method, stabistors (semi-conductor devices which behave in the manner of p–n diodes, conducting heavy current above a certain voltage level) were connected in parallel shunt around each individual nickel–cadmium cell, shunting the charging current from a given cell when the prescribed terminal voltage was reached. This control system is being extended to the AgO/Zn battery. The problem here is that the control voltage is somewhat greater than the 1.5 V breakdown characteristic of the diodes.

Gas Removal. An alternate answer to this gassing problem is to recombine the gas rather than to prevent its formation. Ideally then, the charging circuitry could be simplified and made open-loop. As indicated, oxygen can be removed by providing access of the gas to the negative. Oxygen produced on charging can also be adsorbed on the third electrode, where it is electrochemically reduced to water. An equivalent amount of oxide is formed on the negative [P 14-16,49].

Charging Modes

Regardless of whether open- or closed-loop systems are used, all batteries are charged by one of the methods listed below.

Constant Current.   A constant current is passed through the cell either for a fixed time (open loop) or until the battery is sensed to be charged (closed loop).

Two-Level Constant Current.   A large current is passed until the battery reaches a predetermined state of charge, and then a smaller current is passed to complete the charge.

Exponential Current Charge Rate.   The charging current is continuously reduced from a large initial value toward zero in direct proportion to the remaining amount of charge to be stored. This is the fastest charge method, but is the most difficult to safely instrument, and can be used only on battery cells which can accept the large charge currents initially produced. This method also places the severest strain on any electronic charger.

Constant Potential.   A fixed potential is applied across the battery and held until charge is completed (as evidenced by the battery voltage remaining at the preset charging potential).   This method is usually modified to limit the initial inrush of current into a fully discharged battery. The method is not suitable for battery cells subject to negative resistance effects as they near full charge, since thermal runaway may result.

Combination.   Various combinations of the above methods may be formed to generate essentially any desired current–voltage charging profile.

Charger Circuits

The discussion below is limited to energy converters using the principles of chopper modulation and chopper inversion. A detailed, exhaustive study of circuit techniques applicable to lightweight portable battery chargers capable of operating open-loop or closed-loop, in constant current or constant potential or any programmable combination of these, has been carried out elsewhere [A 320].

The circuit techniques involved chopping of the input power source at a 2-kC rate and passing this AC signal through power control devices operating either as conduction angle, phase control regulators, pulse-width-modulation regulators, or pulse-frequency-modulation regulators. Using the pulse-width control, it was possible to build a charger developing 0–34 V and 0–20 A output at efficiencies in excess of 80% at a charger weight of 30 W/lb for a unit with input source isolation and complete flexibility in the input source (AC or DC, 115 V or 28 V, 40 A or 400 A). Another unit, without source isolation and operating from a DC source of greater potential than the battery potential, developed the same power levels at 120 W/lb. Present semiconductor technology could probably extend the output of this unit by a factor of 4 or 5, providing nearly 3-kW output at a weight from 30 to 100 lb. These output levels are probably sufficient to handle most battery systems requiring the use of nonrotating energy-conversion devices.

No statement has been made of desired charging rates. Generally, the faster the charging rate, the less efficient the overall energy conversion. The maximum rate for solid-state chargers of high efficiency appears to be about 2–3 kW-hr per hour of charging.

Asymmetric Charging Methods

Although batteries can accumulate stored chemical energy only if they are charged with an average DC charging current component, several studies have indicated that inclusion of some AC ripple on the DC charging current improves the physical characteristics of the formed plates.

Apparently, this technique has found some application in the electrodeposition of zinc, nickel, and gold. Superimposed AC greatly improves the adherence of aluminum plated from a bath of $AlCl_3$ and ethyl pyridinium bromide in benzene or toluene [A 281,282].

The application of AC–DC charging to batteries has been much more limited. In one case [A 283], this method was used in the formation of positive Planté plates from sulfuric acid. The peak-to-peak AC voltage was sufficient to cause dissolution of the base metal. The films formed with the superimposed AC were more porous than those formed with DC alone and had lower resistances.

Tests of cell life and its relationship to the gas development showed that, instead of charging with DC, a pulsating rectified AC for recharging increases the life of the battery up to 30%. Apparently, the gas development with its mechanical effects on the plates is much less violent [A 284].

This technique has also been applied to cells containing zinc negatives [P 391] with the result that zinc dendrite growth as well as gas generation is suppressed. A study was made in conjunction with a nickel–zinc system [A 111;P 343] by comparing the weight of dendrites to the total weight of zinc deposited. The electrolyte contained 50 g ZnO/liter KOH (density, 1.4). With a current density of 4 mA/cm², 90% of the zinc was in the form of dendrites after 14 hr. Superimposition of an equal alternating current on the 4 mA/cm² of DC reduced the dendrite formation to zero. The frequency of the AC also exerted an effect. In charging with a single half phase current, dendrite formation was negligible up to 300 cps; at 1000 cps, dendrite formation was 90%. This is consistent with the observations recorded for the aluminum plating experiments mentioned above. With DC charging, the plates were loosely adherent showing dendrites and nodules. However, the superimposed AC yielded a smooth and adherent satin finish.

In another example, the maximum capacity of a AgO/Zn storage battery charged with DC did not surpass 14.3 A-hr, while when charged with AC–DC, capacity reached 16.8 A-hr. The charging wave form had an effect on both the zinc and the silver oxide electrode, providing for a more uniform structure [A 111–113]. The important characteristic of the AC on the positive plate was the duration of the discharging impulse and not the ratio of the duration in both directions. The optimum duration of discharge for batteries charged 10 hr was $\frac{1}{80}$ sec.

A variation on this theme is the use of pulsed charging superimposed on a constant DC charge. A silver oxide electrode was studied under this charging mode [A 285]. Normal DC charges at 100 mA gave a utilization of silver that was generally in the range of 45–53% of the total amount of silver in the electrode, excluding the grid. The maximum utilization measured for charges with added pulses of charging current was 76%, and values from 65 to 70% were obtained frequently when using pulses in 35% KOH. Thus, a 30–40% increase in capacity could be obtained readily by using this method under the proper conditions.

In these experiments, it was noted that the effect occurred primarily during the $Ag_2O \rightarrow AgO$ portion of the charge. "Apparently an increase in charge current increased the potential of the Ag^+ and attracted nearby O^{-2} more strongly. Thus, AgO and possibly Ag_2O were formed deep in the oxide film instead of nearer the surface. It is believed that the expansion of the crystal lattice to form AgO next to the base metal broke the film of oxides which covered the electrode, and allowed electrolyte to penetrate nearer the base metal. Then, O^{-2} no longer had to move through as thick a layer of AgO as before. Thus, polarization was less, following a pulse at the Ag_2O/AgO plateau, than before the pulse occurred. A pulse of high current could be used only for a short period of time before concentration gradients became so large that polarization greatly increased" [A 285].

Dry cells have also been recharged by an AC–DC current. Gas evolution was not observed, nor were zinc dendrites formed [P 415].

Chapter 8

State of the Art—Performance

INTRODUCTION

It is the purpose of this chapter to indicate the general level of performance possible with today's batteries. The data are taken primarily from recent review articles [A 61,365,367] and from manufacturers' literature.

Tables 8-I and 8-II describe the performance of two conventional nonreserve primary battery systems—the LeClanché cell and the Reuben–

TABLE 8-I

Nonreserve Primary Battery Systems (LeClanché Cell)*

Description				
Anode	Zn			
Electrolyte	$NH_4Cl–ZnCl_2$			
Cathode	MnO_2			
Nominal voltage	1.5 V			
Performance				
Operating voltage (average)	1.25–1.35 V			
Energy density				
(W-hr/lb)	20	27	40	67
(W-hr/in.³)	1.6	2.2	3.2	5
Usable temperature range (°F)				
Upper	130	130	113	113
Lower	50	40	40	20
Discharge rate (hr to 0.9 V)	10	30	100	1000

* Taken from [A 365,366].

TABLE 8-II

Nonreserve Primary Battery Systems (Reuben–Mallory Cell)*

Description			
Anode	Zn		
Electrolyte	KOH		
Cathode	HgO		
Nominal voltage	1.34 V		
Performance			
Operating voltage (average)	1.1–1.3 V		
Energy density			
(W-hr/lb)	43–51		
(W-hr/in.^{3})	6.5–7.5		
Usable temperature range (°F)			
Upper	160	160	160
Lower	70	60	35
Discharge rate (hr to 0.9 V)	10	30	100

* Taken from [A 365,366].

Mallory cell, respectively. These systems are not potential high-energy batteries in the sense described in the Preface; the data are included, however, as points of reference.

PRIMARY BATTERIES

Zinc/Silver Oxide

Commercially available batteries generating the highest energy densities are based on the Zn/AgO couple. As mentioned, AgO discharges to silver in two voltage steps. This then can lead to problems in voltage regulation for the battery stack. The other specific problems associated with these reactants have been discussed elsewhere in this volume.

The performance range of primary nonreserve Zn/AgO batteries is shown in Table 8-III. Data for the energy density of a reserve battery system as a function of discharge rate and capacity are given in Table 8-IV. The effective energy density is also a function of the type of activating system. Data are given in Table 8-V for manually activated systems. These energy densities must be reduced at least 50–60% for automatically activated batteries. Manufacturers' performance data for such systems are shown in Table 8-VI.

TABLE 8-III

Nonreserve Primary Battery Systems (Zn/AgO)*

Description	
Anode	Zn
Electrolyte	KOH
Cathode	AgO
Nominal voltage	1.4 V
Performance	
Operating voltage (average)	1.1–1.3 V
Energy density	
(W-hr/lb)	50
(W-hr/in.^{3})	9
Discharge rate (hr to 0.9 V)	10–100

* Taken from [A 365,366].

TABLE 8-IV

Performance of Primary Reserve Zn/AgO Batteries*

Discharge rate	Capacity (A-hr)	Energy density (W-hr/lb)
>200 hr	>15	75–90
1 hr	33	35–45
1 min	5.6	6–10
1 min	0.56	5

* Taken from [A 9,143].

TABLE 8-V

Performance Characteristics for Reserve Zn/AgO Battery
(Manual Activation)*

Operating voltage (average)	1.4–1.5 V			
Energy density				
(W-hr/lb)	35	40	60	75
(W-hr/in.^{3})	2.1	3.5	4.5	6.0
Usable temperature range (°F)				
Upper	165	165	165	125
Lower	80	32	0	−40
Discharge rate (hr to 0.9 V)	$\frac{1}{6}$	1	30	100

* Taken from [A 365,366].

TABLE 8-VI

Reserve Zn/AgO Battery Characteristics (Automatic Activation)*

Capacity (A-hr)	Discharge time (min)	Number of cells	Weight (lb)	Energy density (W-hr/lb)
1	12	20	4.2	6.6
4	12	20	12	9.3
10.6	3.5	22	29.5	10.8
29	12	20	42	19.3
63	20	19	60	41

* Taken from [A 290].

TABLE 8-VII

Estimated Future Possibilities of Zn/AgO System*

Type	Year	Energy density (W-hr/lb)
Nonreserve	1963	50
	1968	60
	1975	70
Reserve	1963	$\leqslant 40$
	1968	70
	1975	100

* Taken from [A 365].

Thus, available Zn/AgO batteries are functioning from 25% to less than 5% of the theoretical capacity of the plate materials. The general reasons for this have been discussed at length. Estimates of future possibilities have been made [A 365] and are summarized in Table 8-VII.

Magnesium Cells

Magnesium is a more active metal than zinc, and, hence, its use in place of zinc should lead to a higher energy density. This anticipated result has not been forthcoming due to the adverse interactions with water described previously. Magnesium electrodes have nevertheless found application in dry cell configurations and, particularly, in sea-water-activated batteries.

The performance of magnesium dry cell batteries is shown in Table 8-VIII. The substitution of m-DNB as the positive material results in the performance figures given in Table 8-IX.

Magnesium batteries used for water activation have generally employed inorganic chlorides as the positive materials, *e.g.*, CuCl and

TABLE 8-VIII

Nonreserve Primary Battery Systems (Magnesium Dry Cell)*

Description				
Anode	Mg			
Electrolyte	$MgBr_2$			
Cathode	MnO_2			
Nominal voltage	1.8 V			
Performance				
Operating voltage (average)	1.4–1.6 V			
Energy density				
(W-hr/lb)	26	40	50	—
(W-hr/in.³)	1.7	2.4	3.5	—
Usable temperature range (°F)				
Upper	130	130	130	130
Lower	50	40	40	20
Discharge rate (hr to 0.9 V)	10	30	100	1000

* Taken from [A 365,366].

TABLE 8-IX

Nonreserve Battery Systems (Mg/m-DNB)*

Description			
Anode	Mg		
Electrolyte	$MgBr_2$ or $Mg(ClO_4)_2$		
Cathode	m-DNB		
Nominal voltage	1.25 V		
Performance			
Operating voltage (average)	1.05–1.15 V		
Energy density			
(W-hr/lb)	60	80	95
(W-hr/in.³)	3.9	4.8	6.7
Discharge rate (hr to 0.9 V)	10	30	100

* Taken from [A 365,366].

TABLE 8-X

Ammonia Cell Components and Initial Performance Levels*

Model	Cell components[†]	Stored liquid component	Peak voltage (V)	Current density (mA/cm²)
NOLC	Mg/KSCN/NH₄SCN–m-DNB–C/SST[‡]	Ammonia	2.0	30
Eglin AFB	Mg/KSCN/NH₄SCN–m-DNB–C/Ag	Ammonia	2.2	0.5
Picatinny A	Mg/KSCN/NH₄SCN–m-DNB–C/SST[‡]	Ammonia	2.0	30 (max.) 1 (min.)
Picatinny C	Mg/KSCN/NH₄SCN–m-DNB–C/Ag	Ammonia	2.1	20
BuWeps	Mg/NH₄SCN/m-DNB–C/SST[‡]	Electrolyte	2.0	100
ECOM	Mg/KSCN/AgCl–C/Ag	Electrolyte	2.0	20
NASA-Lewis	Mg/KSCN/HgSO₄–S–C/Ag	Electrolyte	2.2	1[§]

* Taken from [A ³⁵⁶].

† Cell components listed in the following order: anode, anolyte, cathode.

‡ SST denotes stainless steel.

§ Anode area.

AgCl. Apparently, the most advanced cells (from an engineering standpoint) employ silver chloride. Commercial reserve batteries are available, which, at the 72-hr rate, will produce 70 W-hr/lb (dry) or approximately 40 W-hr/lb (wet) at 5 W-hr/in^3. Activation time is generally less than 10 sec [A 368].

Hydrazine/Oxygen

If this system is operated for extended times, high energy densities are possible. For example, a 6-kW system operated for 42 hr has an energy density of 87 W-hr/lb. However, if discharged for only 2 hr, the energy density is decreased to 25 W-hr/lb. Calculations based on state-of-the-art performance data were also made for a 400-kW, 28-V system operating for 15 min at an ϵ_V of 0.60. The energy density was computed as 8.3 W-hr/lb. At an ϵ_V of 0.50 (the peak power point), this increases to a maximum of 28 W-hr/lb. The limiting factor here is the oxygen electrode.

Unlike the other systems mentioned, the N_2H_4/O_2 battery involves the movement of a fluid and the generation of a gaseous reaction product. These features may influence its final application.

Ammonia-Activated Batteries

A review article on this subject has recently become available [A 356] and is the source of much of the information presented below. The basic chemistry of the cell reaction was discussed in Chapter 3.

"Development of batteries using anhydrous ammonia as the electrolyte solvent has been of great interest for nearly 20 years. At least two desirable properties account for such interest: First, ammonia electrolytes are reasonably conductive and of low viscosity over a wide temperature range; and second, high-energy, high-voltage, light-metal anodes can be used in ammonia electrolytes. These characteristics are particularly attractive for primary reserve battery designs that are required (1) to operate uniformly over a wide temperature range and (2) to activate in a few seconds by remote means without pre-conditioning."

"Investigation of ammonia cells began in 1947, prompted by the discovery of the high-voltage and high-current capabilities of lithium anodes in anhydrous ammonia electrolytes [P 390]. However, battery development under both government and corporate funding was relatively sparse and intermittent throughout this period. The years 1964 and 1965 marked the debut of several ammonia batteries that had undergone sufficient laboratory testing to be regarded as operationally feasible or as ready for pilot production and field testing." These devices are listed in Table 8-X. A comparison of the energy densities of ammonia-activated batteries is given in Table 8-XI. The weight of the activating system was not included, except where noted.

The initial ammonia batteries were activated by the transfer of ammonia vapor to the cell spacer which contained the dry solute. This led to a number of practical difficulties, so that present batteries are now activated with the liquid itself. As indicated in Table 8-X, the complete electrolyte is also used in certain models.

Zinc/Oxygen

Renewed attention [A 369] is being given to the $Zn/KOH/O_2$ system [A 200]. The theoretical energy density of the active plate material is 490 W-hr/lb. This system is being considered for space application employing cryogenic oxygen [A 369]; like fuel cells, the energy density increases with the duration of the mission. "The practical ultimate output, as presently visualized for this system, appears to be in the range 200–250 W-hr/lb" [A 369]. If air is used rather than oxygen in terres-

TABLE 8-XI

Comparison of Energy Densities of Ammonia-Activated Batteries

| Couple | Energy density (W-hr/lb) | | Reference |
	Cell*	System	
Mg/S	44–29		[A 145]
Mg/S†	70		[A 103]
Mg/HgSO$_4$†	53		[A 103]
Mg/HgSO$_4$	26		[A 147]
Mg/AgCl	35	5.6	[A 102,147]
Mg/m-DNB	70‡	10.5	[A 144]

* Cell includes electrodes, electrolyte, and spacer.
† Dual electrolyte cell; −70°C; 3–30-hr rate.
‡ Estimated.

tial applications, it is not necessary to consider the weight of oxygen in sizing the power plant (since oxygen is not stored within the vehicle). In such a case, the theoretical energy density becomes 595 W-hr/lb.

Thermal Batteries

These systems have long shelf lives, since the electrolyte is a solid until power is required. At high temperatures of operation, electrochemical processes are generally quite fast, so that the cells themselves generally can deliver high currents. The Ca/CuO couple has generated 177 W-hr/lb [A 180]. However, when assembled into a battery, the weights of the auxiliaries and activating system decrease the available energy density to less than 10 W-hr/lb. There are additional problems in voltage regulation, which are associated with maintaining a constant battery temperature.

A Li/Cl_2 system is being developed which shows promise of a higher net energy density. A one-cell unit has been operated at 250 W-hr/lb; however, the weights of the heating and insulating systems were not included. By analogy to other thermal batteries, it is expected that a complete Li/Cl_2 system would operate at approximately 20 W-hr/lb for short discharge times. This device is still at the research stage.

Organic Electrolyte Batteries

This work is almost universally concerned with using lithium as a negative. A variety of positives, solvents, and solutes have been surveyed; no truly satisfactory combination has been found.

A Li/PC, $AlCl_3$, LiCl/AgCl, Ag cell, discharged at low current densities (0.23 mA/cm²) yielded energy densities of 28.6–88 W-hr/lb. Cells employing nitromethane as an electrolyte generated 2 W-hr/lb at 4 mA/cm². The energy density figures include the following items: lithium electrodes, silver screen, AgCl mix (90% AgCl), separator, electrolyte, and polyethylene container. Factors not included are: cell case, electrode terminals, insulators, and seals [A 252]. To some extent, this manner of expressing performance does render the data relatively independent of battery size and capacity. The data for other systems presented below also incorporate this convention. It is to be emphasized that care should be exercised in comparing these figures to the conventional performance data for batteries of the AgO/Zn type.

A low current discharge (100-hr rate) was run on the following

TABLE 8-XII

Performance of Secondary Zn/AgO Batteries
Under a Deep Depth of Discharge*

Nominal energy density (W-hr/lb)	Capacity (A-hr)	Discharge time (hr)	Number of cells	Weight (lb)	Cycle life
28.9	100	3	14	69	7
40	165	0.41	18	101	15
36.4	100	16.5	18	79	15
57	21	1.8	16	11	20
27.6	90	0.15	188	800	10
36.8	45	0.22	80	127	10
32	25	0.19	20	18.7	20
23.7	55	0.25	20	65	3

* Taken from [A 290].

TABLE 8-XIII

Performance of Secondary Zn/AgO Batteries*

Capacity output (A-hr)	Nominal energy density (W-hr/lb)	Nominal volume (W-hr/in.³)
100	52	3.2
75	44	3.4
49	45	3.1
25	46	3.1
10	34	2.1
1	38	2.2

* Discharge rate, 60 min; temperature, 70°F. Taken from [A 368].

TABLE 8-XIV

Secondary Zn/AgO Batteries

Depth of discharge (%)	Cycles*
25	300
50	100

* Regime: 90-min discharge and 120-min charge.

cell: $Li/PC–NaPF_6/CuF_2$, Ag. The energy density achieved was 15 W-hr/lb of electrodes and electrolyte. The best cells of this particular series [A 88] used 57–72% of the available CuF_2 in discharging at 1 mA/in.^{2} to a 2.0-V end point. The latter performance is equivalent to 24 W-hr/lb of electrodes and electrolyte.

Improved performance was obtained with Li/BL, (12%) $LiClO_4/CuCl_2$, C [A 85]. In this test cell, two anodes were used per cathode; the weight ratio of $CuCl_2$ to carbon was 3:1; and the separator was a Whatman anion-exchange membrane (AE-30). The cell was discharged at 1 mA/cm^2 for 28 hr from an initial voltage of 3.5 V to a final voltage of 2.8 V. The energy density for the laboratory cell was 106 W-hr/lb. A similar battery was run with a depolarizer of CuF_2 and discharged at 0.5 mA/cm^2 for 30 hr from 3.2 to 3.0 V. The cathode efficiency was 55%; the energy density obtained was 77 W-hr/lb of battery.

A lithium–silver chloride cell [A 120] had an open circuit voltage of approximately 2.9 V, was been discharged at an energy density of 25 W-hr/lb, and has been charge–discharge cycled at up to 10% depths (*i.e.*, <2.5 W-hr/lb).

Cells have been constructed of the system Al/LiCl, $AlCl_3$, EE/AgCl [A 120]. The open circuit cell voltage was 1.4 V; little polarization was obtained at a current density of 2.9 mA/cm^2. The utilization of AgCl was approximately 14%, substantially lower than that obtained with a propylene carbonate electrolyte.

The highest performances reported to date for the Li/CuF_2 system have been 525 W-hr/lb of active material (Li and CuF_2) or 225 W-hr/lb of active material, electrolyte, and support screens [A 376].

SECONDARY BATTERIES

The secondary battery has, as its upper limit, the energy density of the same couple in a primary, nonreserve battery configuration. A result of the requirement of electrochemical reversibility is an increase in weight, *e.g.*, heavier plates are used as well as more extensive separators systems. The dominating inefficiency, however, arises from the limited depth of discharge.

Zn/AgO

The data in Tables 8-XII and 8-XIII describe the performance of some available secondary Zn/AgO batteries under a deep depth of discharge. The depth of discharge is inversely related to the number of cycles (Table 8-XIV).

As indicated previously, the problem of obtaining a larger number of cycles at a higher depth of discharge is related to the amount of soluble zinc released to the electrolyte. A large quantity of zincate leads to negative plate distortion, dendrite growth, and penetration of the separator.

General

Zinc is the most active anode material for use in a secondary battery employing an aqueous electrolyte. An indicated, the magnesium electrode cannot be recharged, presumably due to an insulating oxide–hydroxide film.

The Zn/O_2 system is effectively reversible, and some consideration is being given to its use in as a secondary system. However, little information has been published on this mode of operation.

Lithium can be deposited from LiCl at 600°C and from the low-temperature organic, aprotic solvents. Secondary batteries employing this electrode are still in the research stage.

References

PUBLICATIONS

A 1.   G. Vinal, *Storage Batteries*, John Wiley & Sons (New York), 1955.

A 2.   J. Cotton and I. Dugdale, in: *Batteries*, D. Collins (ed.), MacMillan Co. (New York), 1963, p. 297.

A 3.   J. Duddy and A. Salkind, *J. Electrochem. Soc.* **108**: 717 (1961).

A 4.   H. Francis, *Space Batteries*, NASA SP-5004 (1964), p. 11.

A 5.   S. Kilner and L. Gordon, Contract DA 36-039 SC 567 42, AD 69485 (March 1955); AD 66721 (June 1955); AD 40480 (June 1954).

A 6.   N. Sidgwick, *Chemical Elements*, Oxford University Press (New York), 1950, p. 1101.

A 7.   H. Liebhafsky and E. Oster, Contract DA 44-009-ENG-4909, Report No. 5 (July 1954).

A 8.   W. Latimer, *Oxidation Potentials*, second edition, Prentice-Hall, Inc. (Englewood Cliffs, New Jersey), 1938.

A 9.   D. Colbeck, *Proc. Ann. Power Sources Conf.* **17**: 135 (1963).

A 10.   J. Duddy and A. Salkind, *J. Electrochem. Soc.* **108**: 717 (1961).

A 11.   J. Robinson, *Proc. Ann. Power Sources Conf.* **17**: 142 (1963).

A 12.   H. Knapp and A. Almerini, *Proc. Ann. Power Sources Conf.* **17**: 125 (1963).

A 13.   G. Lozier and R. Ryan, *Proc. Ann. Power Sources Conf.* **16**: 134 (1962).

A 14.   L. Austin, *Trans. Faraday Soc.* **60**: 1319 (1964).

A 15.   J. Newman and C. Tobias, *J. Electrochem. Soc.* **109**: 1183 (1962).

A 16.   H. Lerner and L. Austin, Contract DA 49-186-502-ORD-917, Report No. 6 (May 1964).

A 17.   G. Milazzo, *Electrochemistry*, Elsevier (Amsterdam), 1963, p. 6.

A 18.   S. Zaromb, *J. Electrochem. Soc.* **109**: 1125 (1962).

A 19.   S. Barnartt and D. Forejt, *J. Electrochem. Soc.* **111**: 1201 (1964).

A 20.   *Transcript of the Conference on Secondary Space Batteries* (AD 417060) (1963).

A 21.   R. Jasinski and T. Kirkland, *Mech. Eng.* (March 1964).

A 22.   See [A 59].

A 23.   R. Horne and D. Richardson, *Proc. Ann. Power Sources Conf.* **18**: 75 (1964).

A 24.   E. Settembre and D. Wood, *Proc. Ann. Power Sources Conf.* **16**: 138 (1962).

A 25.   J. Holechek, *Proc. Ann. Power Sources Conf.* **16**: 123 (1962).

A 26.   R. Lewis and A. Partridge, in: *Batteries*, D. Collins (ed.), MacMillan Co. (New York), 1963, p. 171.

A 27.   "*Eveready*" *Battery Applications and Engineering Data Manual*, Union Carbide Consumer Products Co. (New York).

A 28.   J. Rhyne, "Silver Oxide–Zinc Battery Program," WADD Technical Report 61-36 (May 1961).

A 29.   G. Dalin, H. Sulkes, and Z. Strachurski, *Proc. Ann. Power Sources Conf.* **18**: 54 (1964).

A 30.   H. Weiser and M. Roy, *J. Phys. Chem.* **37**: 1009-1018 (1933).

A 31.   C. Wales, *J. Electrochem. Soc.* **111**: 131 (1964).

A 32.   C. Wales, *J. Electrochem. Soc.* **108**: 395 (1961).

A 33.   C. Wales, *J. Electrochem. Soc.* **106**: 885 (1959).

A 34.   T. Dirkse, *J. Electrochem. Soc.* **106**: 453 (1959).

A 35.   R. Amlie and P. Ruetschi, *J. Electrochem. Soc.* **108**: 813 (1961).

A 36.   W. Hammer and D. Craig, *J. Electrochem. Soc.* **104**: 206 (1957).

A 37.   T. Dirkse, *J. Electrochem. Soc.* **109**: 173 (1962).

A 38.   T. Dirkse and L. Vander Lugt, "Study of the Oxides of Silver—No. 4," Contract No. Nonr-1682(01) (June 1957).

A 39.   C. Wales, *J. Electrochem. Soc.* **108**: 395 (1961).

A 40.   C. Wales, *J. Electrochem. Soc.* **109**: 1119 (1962).

A 41.   A. Neiding and I. Kazarnovskii, *Dokl. Akad. Nauk SSSR* **78**: 713 (1951).

A 42.   B. Cahan *et al.*, *J. Electrochem. Soc.* **107**: 725 (1960).

A 43.   T. Dirkse and L. Vander Lugt, *J. Electrochem. Soc.* **111**: 629 (1964).

A 44.   T. Dirkse and F. DeHaan, *J. Electrochem. Soc.* **105**: 311 (1958).

A 45.   T. Boswell, *J. Electrochem. Soc.* **105**: 239 (1958).

A 46.   M. Thompson *et al.*, *J. Electrochem. Soc.* **106**: 737 (1959).

A 47.   M. Straumanis and C. Gill, *J. Electrochem. Soc.* **101**: 10 (1954).

A 48.   B. Roetheli *et al.*, *Metal Alloys* **3**: 73 (1932).

A 49.   U. Evans, *Metallic Corrosion*, Longmans, Green and Co. (New York), 1946, p. 213.

A 50.   R. Glicksman, *J. Electrochem. Soc.* **106**: 458 (1959).

A 51.   G. Akimov and L. Rozenfeld, *Dokl. Akad. Nauk SSSR* **44**: 211 (1944) [C. A. **39**: 4042[4] (1945)].

A 52.   R. Glicksman, *J. Electrochem. Soc.* **106**: 83 (1959).

A 53.   H. Robinson, *Trans. Electrochem. Soc.* **90**: 485 (1946).

A 54.   U. Evans, *op. cit.*, p. 23.

A 55.   H. Robinson, *Corrosion* **2**: 199 (1946).

A 56.   J. Stokes, paper presented at Electrochem. Soc. Meeting (Pittsburgh) (October 1955).

A 57.   E. Schumacher and G. Heise, *J. Electrochem. Soc.* **99**: 191C (1952).

A 58.   G. Akimov and G. Clark, *Trans. Faraday Soc.* **43**: 679 (1947).

A 59.   G. Gruber, *Proc. Ann. Power Sources Conf.* **18**: 94 (May 1964).

A 60.   C. Hampel, *Encyclopedia of Electrochemistry*, Reinhold Publishing Corp. (New York), 1964, p. 77.

A 61.   C. Morehouse, R. Glicksman, and G. Lozier, *Proc. IRE* **46**: 1462 (1958).

A 62.   J. Schrodt, W. Otting, J. Schoegler, and D. Craig, *Trans. Electrochem. Soc.* **90**: 405 (1946).

A 63.   R. Schult and W. Stafford, *Electro-Technol. (New York)* **70**: 84 (June 1961).

A 64.   J. Mullen and P. Howard, *Trans. Electrochem. Soc.* **90**: 529-544 (1946).

A 65.   I. Denison, *Trans. Electrochem. Soc.* **90**: 387-403 (1946).

A 66.   J. White, R. Pierce, and T. Dirkse, *Trans. Electrochem. Soc.* **90**: 467-473 (1946).

A 67.   H. Robinson, *Trans. Electrochem. Soc.* **90**: 485-508 (1946).

A 68.   J. Schrodt, W. Otting, J. Schoegler, and P. Craig, *Trans. Electrochem. Soc.* **90**: 405-417 (1946).

A 69.   B. J. Wilson, Report No. 5998, U. S. Naval Research Lab. (October 4, 1963), PIC No. 823 (February 1964).

A 70.   E. Otto, C. Morehouse, and G. Vinal, *Trans. Electrochem. Soc.* **90**: 419-432 (1946).

A 71.   W. Elliott, S. Hsu, and W. Towle, *Proc. Ann. Power Sources Conf.* **18**: 82 (1964).

A 72.   R. Selian, *Proc. Ann. Power Sources Conf.* **18**: 86 (1964).

A 73.   H. Bauman, *Proc. Ann. Power Sources Conf.* **18**: 89 (1964).

A 74.   J. Farrar, R. Keller, and C. Mazac, *Proc. Ann. Power Sources Conf.* **18**: 92 (1964).

A 75.   M. Eisenberg, paper presented at Electrochem. Soc. Meeting (Washington, D. C.), Abstract No. 46 (October 1964).

A 76.   J. Chilton and R. Holsinger, paper presented at Electrochem. Soc. Meeting (Washington, D. C.), Abstract No. 47 (October 1964).

A 77.   H. Bauman, J. Chilton, and G. Cook, AD 410 577 (July 1963), p. 3.

A 78.   J. Chilton and G. Cook, "New Cathode–Anode Couples Using Non-Aqueous Electrolytes," Report No. ASD-TDR-62-837 (October 1962) (AD 286 899).

A 79.   W. Meyers, "Development of High Energy Density Primary Batteries," Report No. 2, Contract NAS 3-2775 (November 1963) (N 64-16268).

A 80.   R. Horne and D. Richardson, "Low Temperature Operation of Batteries," Report No. 5, Contract No. DA 36-039-SC-90706 (December 1963) (AD 433 871).

A 81.   R. Bird, W. Stewart, and E. Lightfoot, *Transport Phenomena*, John Wiley & Sons (New York), 1960, p. 393.

A 82.   S. Senderoff, E. Klopp, and M. Kronenberg, "High Performance, Short Duration Batteries," Report No. 1, Contract No. NORD-18240 (1958).

A 83.   J. Perry, *Chemical Engineers' Handbook*, McGraw-Hill Book Co. (New York), 1950, p. 474.

A 84.   "Research and Development of a High Capacity Non-Aqueous Secondary Battery," Report No. 4, Contract NAS 3-2780 (1964) (N 64-30863).

A 85.   W. Meyers, Final Report Contract NAS 3-2775 (1964) (N 64-31454).

A 86.   W. Meyers, "Development of High Energy Density Batteries," Report No. 1, Contract NAS 3-6004 (October 1964) (N 64-33617).

A 87.   W. Towle, "Development of High Energy Density Batteries," Report No. 3, Contract No. NAS 3-2790 (April 1964) (N 64-24045).

A 88.   H. Bauman, "Limited Cycle Secondary Battery Using Lithium Anode," Report No. APL-TDR-64-59 (May 1964) (AD 601 128).

A 89.   J. Chilton, "New Cathode–Anode Couples," Report No. ASD-TDR-62-1 (April 1962) (AD 277 171).

A 90.   F. Hurley and T. Weir, *J. Electrochem. Soc.* **98**: 203 (1951).

A 91.   A. Lyall and H. Seiger, "Lithium–Nickel Halide Secondary Battery," Report No. 1, Contract AF 33(615)-1266 (March 1964) (AD 433 616).

A 92.   A. Lyall and H. Seiger, Report No. 3, Contract AF 33(615)-1266 (September 1964) (AD 605 754).

A 93.   H. Bauman, J. Chilton, and G. Cook, "New Cathode–Anode Couples," Report No. 1, Contract AF 33(616)7957 (December 1962) (AD 294 308).

A 94.   H. Bauman, "Lithium Anode Limited Cycle Secondary Battery," Report No. 3, Contract AF 33(657)11709 (March 1964) (AD 433 543).

A 95.   J. Chilton, W. Conner, and R. Holsinger, "Lithium–Silver Chloride Secondary Battery Investigation," Report No. 1, Contract AF 33(615) 1195 (April 1964) (AD 439 399).

A 96.   C. Hodgman, *Handbook of Chemistry and Physics*, 44th edition, Chemical Rubber Co. (Cleveland), 1962, p. 1749.

A 97.   L. J. Minnick, *Proc. Ann. Power Sources Conf.* **14**: 114-116 (1960).

A 98.   G. F. Sieglinger, *Proc. Ann. Power Sources Conf.* **15**: 93-95 (1961).

A 99.   J. M. Freund and R. W. Graham, "Basic Research for Ammonia Vapor Activated Batteries," Report No. 1, Contract DA 36-039-sc-72306 (July 1956) (AD 119827).

A 100. O. Adlhart, *Proc. Ann. Power Sources Conf.* **14**: 111-113 (1960).

A 101. L. R. Wood and D. J. Doan, *Proc. Ann. Power Sources Conf.* **17**: 132-134 (1963).

A 102. D. J. Doan and L. R. Wood, "Research on Ammonia Battery System," Report No. 12, Contract DA 36-039-sc-89188 (July 1963) (AD 419401).

A 103. W. F. Meyers *et al.*, "Development of High Energy Density Primary Batteries," Report No. 1, Contract NAS3-2775 (August 1963) (N64-10901).

A 104. W. S. Harris and G. B. Matson, "The Electroreduction of Aromatic Nitro Compounds in Nonaqueous Solutions; Part I. The Reduction of Nitrobenzenes in Acid–Liquid Ammonia Solutions," Report No. NAVWEPS 7241 (December 1962) (AD 293446).

A 105. W. S. Wong, "Sixth Symposium on Ammonia Batteries," Report No. NOLC 597 (January 1964) (AD 433 973).

A 106. F. E. Rosztoczy, "Fourth Symposium on Ammonia Batteries," Report No. NOLC 559 (January 1962) (AD 272289).

A 107. W. L. Jolly, "Metal–Ammonia Solutions," in: *Progress in Inorganic Chemistry*, *Vol. 1*, A. Cotton (ed.), Interscience (New York), 1959, pp. 235-282.

A 108. H. H. Sisler, *Chemistry in Non-Aqueous Solvents*, Reinhold (New York), 1961.

A 109. L. F. Audreth and J. Kleinberg, *Non-Aqueous Solvents*, second edition, John Wiley & Sons (New York), 1965.

A 110. G. Jander, "Die Chemie in Wasserähnlichen Lösungsmitteln," Chapter 3, in: *Chemistry in Non-Aqueous Solvents*, H. H. Sisler (ed.), Reinhold (New York), 1961.

A 111. V. Romanov, *Zh. Prikl. Khim.* **34**: 2692 (1961) (C.A. **56**: 9866 d).

A 112. V. Romanov, *Vestn. Elektroprom.* **31**: 26 (1960) (C.A. **55**: 6197 i).

A 113. V. Romanov, *Zh. Prikl. Khim.* **34**: 1312 (1961) (C.A. **55**: 2311 g).

A 114. W. Wawrzyozek, *Przemysl Chem.* **38**: 479 (1959) (C.A. **54**: 13568 g).

A 115. W. Wawrzyozek, *Zh. Fiz. Khim.* **35**: 723 (1961) (C.A. **55**: 16054 e).

A 116. F. Kukoz and L. Kukoz, *Tr. Novocherk. Politekhn. Inst.* **79**: 207 (1959) (C.A. **54**: 2400 i).

A 117. A. Trofimov, *Primenenie Ul'traakustiki k Issled. Veshchesta*, No. 10, 1960, p. 103 (C.A. **55**: 20722 c).

A 118. L. Kukoz and M. Skalozubov, *Tr. Novocherk. Politekhn. Inst.* **134**: 13–30 (1962) (C.A. **58**: 12163 a).

A 119. J. Farrar, R. Keller, and C. Mazac, "High Energy Battery System Study," Report No. 4, Contract DA-36-039-AMC-03201(E) (June 1964) (AD 450559).

A 120. J. Chilton, W. Conner, and R. Holsinger, see [A 95], Report No. 2 (July 1964) (AD 450428).

A 121. N. Sidgwick, *Chemical Elements*, Oxford University Press (New York), 1950, p. 101.

A 122. S. Homan, *Physico-Chemical Effects of Pressure*, Butterworths Scientific Publ. (London), 1957, p. 126.

A 123. R. Horne, *J. Electrochem. Soc.* **110**: 1282 (1963).

A 124. S. B. Brummer, *J. Chem. Phys.* **41**: 1636 (1965).

A 125. R. Robinson and R. Stokes, *Electrolytic Solutions*, Butterworths Scientific Publ. (London), 1959, p. 288.

A 126. D. MacInnes, *Principles of Electrochemistry*, Reinhold Publ. Co. (New York), 1939, p. 116.

A 127. J. Farrar *et al.*, see [A 119], Report No. 1 (September 1963) (AD 429290).

A 128. J. Chilton, see [A 89] (October 1963) (AD 425876).

A 129. H. Bauman, see [A 94], Report No. 2 (December 1963) (AD 426791).

A 130. J. Chilton, see [A 89], Report No. 3 (July 1963) (AD 411314).

A 131. D. Chrisafulli, "A New Cycle for Ni/Cd Cells," Contract NAS 52366 (February 1965).

A 132. R. Panzer, "Chemoelectric Energy Conversion," NAVWEPS Report 8210 (September 1964) (AD 451296).

A 133. R. Muller, paper presented at Fourth Symposium on Ammonia Batteries (1962) (AD 272289).

A 134. R. Meredith, paper presented at Fourth Symposium on Ammonia Batteries (1962) (AD 272289).

A 135. D. Doan, "Research on Ammonia Battery Systems," Report No. 6, Contract No. DA-36-039-sc-85396 (January 1962) (AD 274381).

A 136. R. Foley, "High Rate Experimental Ammonia Battery," Contract No. NOW 64-0568-d (September 1964).

A 137. M. Wadley, "The HF Solvent System," Thesis No. 64-5772 (1963).

A 138. A. Clifford and E. Zamora, *Trans. Faraday Soc.* **57**: 1963 (1961).

A 139. J. Simons and H. Hildebrand, *J. Am. Chem. Soc.* **46**: 2223 (1924).

A 140. S. Kongpricha, M.Sc. Thesis, Purdue University (1956).

A 141. E. Schaschl, *J. Electrochem. Soc.* **94**: 299 (1948).

A 142. M. Schaer and R. Meredith, Sixth Symposium on Ammonia Batteries (January 1964) (AD 433973), p. 24.

A 143. A. Hellfritzsch, paper presented at SAE Meeting (New York) (September 1964).

A 144. L. Minnick, *Proc. Ann. Power Sources Conf.* **17**: 129 (May 1963).

A 145. V. Bryant, J. Freund, and J. Schliff, "High Energy Batteries," Contract No. NOrd-18249, Final Report (June 1961) (AD 260700).

A 146. D. Doan and L. Wood, see [A 102], Report No. 15 (March 1964) (AD 447190).

A 147. A. Almerini, Sixth Symposium on Ammonia Batteries (January 1964) (AD 433973).

A 148. *Handbook of Chemistry and Physics*, 45th edition, Chemical Rubber Co. (Cleveland), 1964.

A 149. J. Timmermans, *Physico-Chemical Constants of Pure Organic Compounds*, Elsevier (New York), 1950.

A 150. R. Robinson and R. Stokes, *Electrolytic Solutions*, Butterworths Scientific Publ. (London), 1959, p. 10.

A 151. W. Harris, Thesis, Report UCRL 8381 (July 1958).

A 152. H. Laitinen and H. Gaur, *Anal. Chim. Acta* **18**: 1 (1958).

A 153. R. Parsons, *Handbook of Electrochemical Constants*, Butterworths Scientific Publ. (London), 1959, p. 79.

A 154. T. Reddy, *Electrochem. Tech.* **1**: 325 (1963).

A 155. C. Liu, *J. Phys. Chem.* **66**: 164 (1962).

A 156. H. Swofford and H. Laitinen, *J. Electrochem. Soc.* **110**: 814 (1963).

A 157. I. Delimarskii and B. Markov, *Electrochemistry of Fused Salts*, Sigma Press Publ. (Washington, D.C.), 1961.

A 158. D. Ives and G. Janz, *Reference Electrodes*, Academic Press (New York), 1961.

A 159. S. Selis, J. Wondawski, and R. Justus, *J. Electrochem. Soc.* **111**: 6 (1964).

A 160. R. Goodrich and R. Evans, *J. Electrochem. Soc.* **99**: 207C (1952).

A 161. H. Laitinen and B. Bhatia, *J. Electrochem. Soc.* **107**: 705 (1960).

A 162. S. Selis, L. McGinnis, E. McKee, and J. Smith, *J. Electrochem. Soc.* **110**: 470 (1963).

A 163. S. Selis and L. McGinnis, *J. Electrochem. Soc.* **106**: 900 (1959).

A 164. S. Selis and L. McGinnis, *J. Electrochem. Soc.* **108**: 191 (1961).

A 165. S. Selis, G. Elliott, and L. McGinnis, *J. Electrochem. Soc.* **106**: 134 (1959).

A 166. H. Laitinen and D. Rhodes, *J. Electrochem. Soc.* **109**: 413 (1962).

A 167. H. Laitinen and C. Liu, *J. Am. Chem. Soc.* **80**: 1015 (1958).

A 168. E. Van Artsdalen and I. Yaffe, *J. Phys. Chem.* **59**: 118 (1955).

A 169. H. Laitinen and J. Pankey, *J. Am. Chem. Soc.* **81**: 1053 (1959).

A 170. S. Flengas and T. Ingraham, *J. Electrochem. Soc.* **106**: 714 (1959).

A 171. E. Klopp and S. Senderoff, Second Report NOrd 19240 (December 1958) (AD 226262).

A 172. E. Uhler, D. Fields, and G. Lozier, Second Report AF 33(657)7758 (April 1962) (AD 275524).

A 173. J. Fuscue, S. Carlton, and D. Levertz, WADD Report 60-442 (1961).

A 174. R. Weaver, *Proc. Ann. Power Sources Conf.* **19**: 113 (May 1965).

A 175. J. Kennedy, R. Eliason, and J. Adams, 149th ACS Meeting (Detroit, Michigan) (April 1965).

A 176. A. Wheeler, in: *Catalysis, Vol. 3*, P. Emmett (ed.), Reinhold Publ. (New York), 1955, p. 123.

A 177. M. Indig and R. Snyder, *J. Electrochem. Soc.* **109**: 757 (1962).

A 178. S. Yoshizawa, Z. Takehara, and H. Katsuya, *Denki Kagaku* **32**: 519 (1964) [*C.A.* **62**: 10062 h (1965)].

A 179. V. Flerov, *Zh. Prikl. Khim.* **34**: 1929 (1961).

A 180. S. Senderoff, E. Klopp, and M. Kronenberg, Final Report NOrd 18240 (June 1962) (AD 277433).

A 181. B. Larrick, U.S.N.O.L., PIC 796 (April 1964).

A 182. E. McKee, *Proc. Ann. Power Sources Conf.* **10**: 26 (May 1956).

A 183. W. Ferguson, Thesis, University of Illinois, Urbana (1956).

A 184. C. Jennings, *J. Electrochem. Soc.* **103**: 531 (1956).

A 185. H. Swofford and H. Laitinen, *J. Electrochem. Soc.* **110**: 814 (1963).

A 186. K. Johnson and H. Laitinen, *J. Electrochem. Soc.* **110**: 314 (1963).

A 187. H. Gaur and W. Bahl, *Electrochim. Acta* **8**: 107 (1963).

A 188. K. Grjotheim, *Z. Physik. Chem. (Frankfurt)* **11**: 150 (1957).

A 189. E. Uhler, J. Eigen, and G. Lozier, Contract AF 33(616)7505 (AD 268267) (1961); (AD 275524) (1962); (AD 284061) (1962).

A 190. W. Subkasky *et al.*, Second Quart. Report NASA CR 54289 (March 1965) (N65-15882).

A 191. G. Zellhoefer, U.S. 3, 110, 632 (November 1963).

A 192. J. Robinson and P. King, *J. Electrochem. Soc.* **108**: 36 (1961).

A 193. A. Kuzmina and L. Kuzmin, *Zh. Prikl. Khim.* **36**: 335 (1963).

A 194. S. Kilner *et al.*, Final Report, Contract DA-36-039-SC-71209 (April 1957) (AD-138363).

A 195. L. McGraw *et al.*, Report No. 1, Contract DA-36-039-SC-42682 (June 1953) (AD 14500).

A 196. *Ibid.*, Report No. 2 (September 1953) (AD-24177).

A 197. *Ibid.*, Report No. 3 (1).

A 198. R. Jasinski, *J. Electrochem. Soc.* **112**: 526 (1956).

A 199. D. Ives and G. Janz, *Reference Electrodes*, Academic Press (New York), 1961.

A 200. G. W. Vinal, *Primary Batteries*, John Wiley & Sons (New York), 1950, p. 284.

A 201. *Ibid.*, p. 279.

A 202. J. Popova, *Zh. Prikl. Khim.* **36**: 1687 (1963).

A 203. R. Shair *et al.*, Final Report, Contract NAS 5-809 (July 1963) (N65-17518).

A 204. A. Charkey and G. Dalin, Report No. 4, Contract NAS 5-3452 (June 1964) (N 64-33187).

A 205. General Electric Co., Report No. 2, Contract NAS 5-3669 (November 1964) (N 65-19754).

A 206. A. Fleischer, *Proc. Ann. Power Sources Conf.* **13**: 83 (1959).

A 207. F. Cushing, Report No. 1, Contract NAS 5-3813 (September 1964) (N 64-33918).

A 208. A. Catotti, *Proc. Ann. Power Sources Conf.* **19**: 63 (May 1965).

A 209. M. Sulkes and G. Dalin, Report No. 4, Contract No. DA-36-039-AMC-02238(E) (June 1964) (N 65-15759).

A 210. *Ibid.*

A 211. Symposium, *Elektrokhim.* **1**: 381 (1965).

A 212. V. Flerov, *Zh. Prikl. Khim.* **34**: 1929 (1961).

A 213. J. Lander and J. Keralla, Report No. 2, Contract AF33(657)8943 (January 1963) (N 65-13084).

A 214. V. Flerov, *Izv. Vysshykh. Uchebn. Zavedenii Khim. i Khim. Tekhnol.* **6**: 829 (1963) (C.A. **60**: 11610 c).

A 215. G. Mikhailenko and M. Skalozubov, *Zh. Prikl. Khim.* **38**: 420 (1965).

A 216. W. Pauli, Contract DA-5 NO6-01-010 (March 1961) (AD-253177).

A 217. A. Salkind and P. Bruins, *J. Electrochem. Soc.* **109**: 356 (1962).

A 218. H. Ritterman, Report No. 1, Contract NAS-34178 (September 1964) (N65-10047).

A 219. B. Conway and P. Bourgault, *Can. J. Chem.* **38**: 1557 (1960).

A 220. G. Briggs and W. Wynne-Jones, *Trans. Faraday Soc.* **52**: 1272 (1956).

A 221. G. Briggs and W. Wynne-Jones, *Electrochim. Acta* **7**: 241 (1962).

A 222. P. Lukovetsev and G. YaSlaidin, *Zh. Fiz. Khim.* **36**: 2268 (1962).

A 223. C. Nordell, Report ECO M-2517 (1965) (AD 609988).

A 224. E. Kantner *et al.*, Report No. 1, Contract AF 33(615)-2087 (November 1964) (N65-13072).

A 225. E. Butler, Report No. 32-535 JPL. (December 1963) (N 64-13285).

A 226. H. Johnston, F. Çata, and A. Garrett, *J. Am. Chem. Soc.* **55**: 2311 (1933).

A 227. Y. Pleskov and B. Kakanov, *Zh. Neorgan. Khim.* **2**: 1807 (1957).

A 228. T. Dirkse and B. Wiers, *J. Electrochem. Soc.* **106**: 284 (1959).

A 229. A. I. Oshe *et al.*, *Zh. Prikl. Khim.* **34**: 2254 (1961).

A 230. Z. Stachurski and G. Dalin, Report No. 1, Contract NAS 5-3873 (July 1964) (N 65-13526).

A 231. *Ibid.*, Report No. 2 (December 1964) (N65-22657).

A 232. G. Dalin and M. Sulkes, *Proc. Ann. Power Sources Conf.* **19**: 69 (May 1965).

A 233. See [A 204].

A 234. E. Weiss, Report No. 8, Contract NAS 5-2860 (July 1964) (N 65-11631).

A 235. J. Mackie and P. Meares, *Proc. Roy. Soc.* (*London*) **A 232**: 498, 509 (1955).

A 236. G. Dalin and Z. Stachurski, Paper No. 244, Electrochem. Soc. Meeting (San Francisco) (May 1965).

A 237. R. Horne and D. Richardson, Final Report, Contract DA-36-039-SC-90706 (July 1964) (N 65-12752).

A 238. G. Lago *et al.*, Final Report, Contract DA-36-039-SC-74994 (March 1959) (PB 143899).

A 239. M. Arcand, Final Report, Contract DA 36-039-SC-72363 (July 1957) (PB 135749).

A 240. G. Lozier *et al.*, Report No. 7, Contract DA-36-039-SC-85340 (February 1962).

A 241. S. Naiditch, Report No. 1, Contract NAS 7-326 (November 1964) (N65-15956).

A 242. I. Kolthoff and J. Coetzee, *J. Am. Chem. Soc.* **79**: 870, 1852 (1957).

A 243. A. Popov and D. Geske, *J. Am. Chem. Soc.* **79**: 2074 (1957).

A 244. I. V. Nelson, Thesis No. 64-8509 (June 1963).

A 245. V. Gutman and G. Schober, *Z. Anal. Chem.* **171**: 339 (1959).

A 246. I. Kolthoff and T. Reddy, *J. Electrochem. Soc.* **108**: 980 (1961).

A 247. G. Brown and R. Al-Urfali, *J. Am. Chem. Soc.* **80**: 2113 (1958).

A 248. J. Chilton *et al.*, Report No. 3, Contract AF 33(615)-1195 (October 1964) (AD 607968).

A 249. W. Elliott *et al.*, Report No. 1, Contract NAS 3-6015 (September 1964) (N 65-11518).

A 250. P. R. Mallory & Co., Report No. 1, Contract NAS 3-6017 (April 1965).

A 251. J. Farrar *et al.*, Report No. 5, Contract DA 36-039-AMC-03201(E) (September 1964) (AD 458472).

A 252. J. Chilton *et al.*, Report No. AFAPL-TR-64-147 (February 1965) (AD 612189).

A 253. R. Panzer, Paper No. 164, Electrochem. Soc. Meeting (San Francisco) (May 1965).

A 254. W. Meyers, Report No. 3, Contract NAS 3-6009 (March 1965).

A 255. J. Daley, G. McWilliams, and W. Spindler, Navweps Report 8813, NOL, Corona, California (April 1965).

A 256. D. Faletti and L. Nelson, *Electrochem. Tech.* **3**: 98 (1965).

A 257. D. Doan, *Proc. Ann. Power Sources Conf.* **11**: 64 (May 1957).

A 258. E. McKee and J. Smith, Final Report, Contract DA-36-039-SC-42602 (1955); Contract DA 36-039-SC-64617 (1958).

A 259. R. Panzer, Tech. Memo 44-7 USNOL, Corona, California (1960).

A 260. O. Stafford, *J. Am. Chem. Soc.* **55**: 3987 (1933).

A 261. A. Dunstan and A. Mussell, *J. Chem. Soc.* **97**: 1935 (1910).

A 262. G. Jander and G. Winkler, *J. Inorg. Nucl. Chem.* **9**: 24 (1959).

A 263. L. Yntema and L. Audrieth, *J. Am. Chem. Soc.* **52**: 2693 (1930).

A 264. R. Wallace, Thesis No. 64-330 (1963).

A 265. W. Elliott *et al.*, Final Report, Contract NAS 3-2790 (January 1965).

A 266. A. Lyall *et al.*, Report AFAPL-TR-65-11 (January 1965) (N65-20555).

A 267. Y. Yokota, *Denki Kagaku* **28**: 518 (1960) (C.A. **62**: 2503 b).

A 268. I. Rozenfeld and T. Lukonina, *Dokl. Akad. Nauk SSSR* **111**: 136 (1956) (C.A. **52**: 3564 a).

A 269. J. Coleman, *Trans. Electrochem. Soc.* **93**: 545 (1946).

A 270. J. Coleman, *J. Electrochem. Soc.* **98**: 26 (1951).

A 271. E. Grens and C. Tobias, *Ber. Bansenges* **68**: 236 (1964).

A 272. E. Grens and C. Tobias, Report NASA-CR-59573 (1965).

A 273. P. Lukovtsev, *Tr. 4-go. Soveshch. Elektrokhim. (Moscow)* (1956), p. 773 (C.A. **54**: 9543 f).

A 274. C. Shepherd, *J. Electrochem. Soc.* **112**: 657 (1965).

A 275. J. Wallinder, *Lead–Acid Batteries*, Lead Development Association (London), March 1965.

A 276. R. Wrabetz, *Z. Elektrochem.* **60**: 722 (1956).

A 277. R. Golding, *J. Chem. Soc.* **1959**: 1838.

A 278. M. Iwasaki *et al.*, *J. Electrochem. Soc. Japan* **22**: 578 (1954).

A 279. D. Yamashita and Y. Yamamoto, *Denki Kagaku* **32**: 684 (1964) (C.A. **62**: 8663 a).

A 280. L. Antipin, *Zh. Fiz. Khim.* **30**: 1425 (1956) (C.A. **51**: 6394 i).

A 281. F. Hurley and T. Wier, *J. Electrochem. Soc.* **98**: 207 (1951).

A 282. W. H. Safranek, W. Schickner, and C. Faust, *J. Electrochem. Soc.* **99**: 53 (1952).

A 283. R. Vijayaralli *et al.*, *J. Electrochem. Soc.* **110**: 1 (1963).

A 284. F. Dacos, *Rev. Universelle Mines* **2**: 3 (1946) (C.A. **40**: 6347[6]).

A 285. C. Wales, *J. Electrochem. Soc.* **111**: 131 (1964).

A 286. E. Uhler *et al.*, Final Report, Contract No. DA 36-039-SC-78048 (1960) (AD 403933).

A 287. R. Glicksman and C. Morehouse, *J. Electrochem. Soc.* **105**: 299 (1958).

A 288. B. Gruber *et al.*, Final Report, Contract No. DA 36-039-SC-87336 (1964) (AD 454 913).

A 289. L. Fischer, G. Winkler, and G. Jander, *Z. Elektrochem.* **62**: 1 (1958).

A 290. Electric Storage Battery Co., Missile Battery Division, Raleigh, N.C.

A 291. *Engineering Manual*, Burgess Battery Co. (Freeport, Illinois), 1964.

A 292. J. Jones and A. Arranaga, *J. Electrochem. Soc.* **105**: 435 (1958).

A 293. P. King, *J. Electrochem. Soc.* **110**: 1113 (1963).

A 294. A. Tripler and L. McGraw, *J. Electrochem. Soc.* **105**: 179 (1958).

A 295. S. Yoshizawa and Z. Takehara, *Electrochim. Acta* **5**: 240 (1961).

A 296. J. Cooper and A. Fleischer (eds.), "Battery Separator Screening Methods," Air Force Aero Propulsion Lab., Wright-Patterson AFB (1964) (AD 447301).

A 297. T. Hoar, in: *Modern Aspects of Electrochemistry, Vol. 2*, J. O'M Bockris (ed.), Butterworths Scientific Publ. (London), 1959.

A 298. J. A. Butler, *Electrocapillarity*, Methuen Publ. (London), 1940.

A 299. M. Pourbaix, N. Zoubov, and J. Van Muylder, *Atlas D'Equilibres Electrochimiques*, Gauther-Villars & Co. (Paris), 1963.

A 300. A. Parker, *Science and Technology*, No. 44 (August 1965).

A 301. W. Elliott *et al.*, 3rd Quarterly Report, Contract NAS 3-6015 (March 1965).

A 302. E. Gillis, ERDL, Ft. Belvoir, Virginia, personal communication (1964).

A 303. Panel discussion, "Nature of AgO," Electrochem. Soc. Meeting (Buffalo, N.Y.) (October 1965).

A 304. A. Kozawa and J. Yeager, *J. Electrochem. Soc.* **112**: 176C (1955).

A 305. F. Kober, *J. Electrochem. Soc.* **112**: 172C (1965).

A 306. W. Carson, *Proc. Ann. Power Sources Conf.* **18**: 59 (May 1964).

A 307. H. Seiger, R. Shair, and P. Ritterman, *Proc. Ann. Power Sources Conf.* **18**: 61 (May 1964).

A 308. F. Cocca, *Proc. Ann. Power Sources Conf.* **18**: 65 (May 1964).

A 309. M. Schaer and R. Meredith, *J. Electrochem. Soc.* **112**: 1746 (1965).

A 310. S. Glasstone, *Textbook of Physical Chemistry*, second edition, D. Van Nostrand and Co. (New York), 1946, p. 904.

A 311. W. Carson, Contract NAS 5-3669 (March 1965).

A 312. G. Cohen and E. Ostrander, Report No. LTR 64-13 (June 1964) (N 65-18659).

A 313. M. Sulkes and G. Dalin, Report No. 6, Contract No. DA 36-039-AMC-02238(E) (December 1964) (AD 464470).

A 314. P. Allen and A. Hickling, *Trans. Faraday Soc.* **53**: 1626 (1957).

A 315. T. Dirkse and J. DeRoos, *Z. Physik. Chem. (Frankfurt)* **41**: 1 (1964).

A 316. W. Bishop, APIP Tech. Memorandum 64-2 (N 65-18382).

A 317. H. Francis, NASA Report SP 5004 (1964).

A 318. P. Benson, G. Briggs, and W. Wynne-Jones, *Electrochim. Acta* **9**: 275 (1964).

A 319. *Ibid.*, p. 281.

A 320. R. Craven and C. Moo, Report No. 2, Contract DA-36-039-AMC-02330(E) (December 1964) (AD-600044).

A 321. H. Gerischer, *Anal. Chem.* **31**: 33 (1959).

A 322. P. Delahay, *Double Layer and Electrode Kinetics*, Interscience (New York), 1965, p. 157.

A 323. T. Berzins and P. Delahay, *Z. Elektrochem.* **59**: 792 (1955).

A 324. R. Birke and D. Roe, *Anal. Chem.* **37**: 450 (1965).

A 325. K. Vetter, *Electrochemische Kinetik*, Springer-Verlag (Berlin), 1961, p. 263.

A 326. M. Fleischman and M. Thirsk, *Electrochim. Acta* **2**: 22 (1960).

A 327. M. Fleischman and M. Thirsk, in: *Advances in Electrochemistry*, P. Delahay (ed.), Interscience (New York), 1963, p. 123.

A 328. J. Bockris and A. Damjanovic, *Modern Aspects of Electrochemistry*, Butterworths Scientific Publ. (London), 1964, p. 244.

A 329. B. Conway and J. Bockris, *Electrochim. Acta* **3**: 340 (1961).

A 330. H. Gerischer, *Z. Elektrochem.* **62**: 256 (1958).

A 331. W. Burton, W. Cabrara, and F. Franks, *Nature* **163**: 398 (1949).

A 332. D. Vermilyea, *J. Chem. Phys.* **27**: 814 (1957).

A 333. R. Kaishev, E. Budevski, and Ya. Malinovski, *Z. Physik. Chem. (Leipzig)* **204**: 348 (1955).

A 334. H. Pick, *Nature* **176**: 693 (1955).

A 335. H. Seiter, H. Fischer, and C. Albert, *Electrochim. Acta* **2**: 167 (1960).

A 336. M. Fleishman and M. Thirsk, in: *Advances in Electrochemistry*, P. Delahay (ed.), Interscience (New York), 1963, p. 156.

A 337. M. Fleischman and M. Thirsk, *Electrochim. Acta* **1**: 146 (1959).

A 338. M. Fleischman and M. Liler, *Trans. Faraday Soc.* **54**: 1370 (1958).

A 339. M. Fleischman and M. Thirsk, *Trans. Faraday Soc.* **51**: 71 (1955).

A 340. G. Wranglen, *Electrochim. Acta* **2**: 130 (1960).

A 341. N. Pangaroff, *Electrochim. Acta* **7**: 139 (1962).

A 342. H. Fischer, *Electrochim. Acta* **2**: 50 (1960).

A 343. R. Burke, Thesis, M.I.T. (September 1965).

A 344. I. Sanghi and M. Fleischman, *Electrochim. Acta* **1**: 101 (1959).

A 345. C. Wales and J. Burbank, *J. Electrochem. Soc.* **112**: 13 (1965).

A 346. V. Scatturin, P. Bellon, and A. Salkind, *J. Electrochem. Soc.* **108**: 819 (1961).

A 347. A. McKie and D. Clark, *Batteries*, Pergamon Press (New York), 1963, p. 285.

A 348. Ts. Yu Vol and N. Shishakov, *Izv. Akad. Nauk SSSR Otd. Khim. Nauk* (1962), p. 586.

A 349. C. Carvajal, K. Tolle, J. Smid, and M. Szwarc, *J. Am. Chem. Soc.* **87**: 5548 (1965).

A 350. A. Stearn and H. Eyring, *J. Chem. Phys.* **5**: 113 (1937).

A 351. S. Brummer and G. Hills, *Trans. Faraday Soc.* **57**: 1816, 1823 (1961).

A 352. V. Sochevanov, *J. Gen. Chem. USSR (Eng. Transl.)* **22**: 1119 (1952) (C.A. **40**: 31296).

A 353. T. Dirkse, *J. Electrochem. Soc.* **102**: 497 (1955).

A 354. J. Rile, ASD-TDR-63-560 (August 1963) (AD 418143).

A 355. W. Meyers and G. Armstrong, Final Report NASA CR 54733 (June 1965).

A 356. W. Spindler and R. Pritchard, paper presented at Advances in Battery Technology Symposium (Los Angeles) (1965).

A 357. J. Byrne *et al.*, Quart. Rept. No. 1, NASA-CR-54782 (September 1965).

A 358. W. Elliott *et al.*, see [A 301], 5th Quarterly Report (September 1965).

A 359. I. Gillet, *J. Electrochem. Soc.* **108**: 71 (1961).

A 360. R. Amlie, H. Honer, and P. Ruetschi, *J. Electrochem. Soc.* **112**: 1073 (1965).

A 361. D. Wood, ECOM 2561 (February 1965) (AD 615114).

A 362. J. Pawlak, ECOM 2561 (March 1965) (AD 616120).

A 363. G. Thaller, NASA TND 2915 (July 1965).

A 364. Anon., Report No. 1, DB 80685 (October 1943).

A 365. A. Fleischer, in: *Bgd. Material for the Study of the National Space Power Program, Vol. 1*, Power Inf. Agency (Philadelphia) (November 1964) (AD 452617).

A 366. J. Hovendon, Power Sources Branch, Components Dept., USAERDL, Fort Monmouth, N.J.

A 367. R. Garner, Report RE-TR-64-29 (December 1964) (N65-15907).

A 368. *Energy Data Book*, Yardney Electric Corp. (New York).

A 369. G. Starkey, "Introduction to Session on Secondary Batteries," *Proc. Ann. Power Sources Conf.* **19** (May 1965).

A 370. H. Sand, *Phil. Mag.* **1**: 45 (1901).

A 371. R. Murray and C. Reilley, *Electroanalytical Principles*, Interscience (New York), 1963.

A 372. M. Klochko and M. Godneva, *J. Inorg. Chem.* (*USSR*) **4**: 2127 (1959).

A 373. *International Critical Tables*, *Vol. IV*, McGraw-Hill (New York), 1928, p. 239.

A 374. L. Darken and H. Meier, *J. Am. Chem. Soc.* **64**: 621 (1942).

A 375. C. Shepherd and H. Langelan, *J. Electrochem. Soc.* **109**: 657 (1962).

A 376. W. Robertson, NASA–Lewis Research Center, personal communication (January 1965); see also reports of Livingston Electr. Corp.

A 377. B. Conway, *Electrode Processes*, Ronald Press (New York), 1965.

PATENTS

P 1	G. Neuman	US 2,571,927	October 1951
P 2	P. Jacquier	US 2,614,138	October 1952
P 3	F. Peters	US 2,798,110	July 1957
P 4	W. Germershausen	US 2,842,607	July 1958
P 5	F. Peters	US 2,988,584	June 1961
P 6	F. Peters	US 3,031,517	April 1964
P 7	M. Strauss	US 3,057,943	October 1962
P 8	H. Vogt	US 3,059,041	October 1962
P 9	J. Duddy	US 2,968,686	January 1961
P 10	H. Usel	US 3,000,996	September 1961
P 11	M. Strauss	US 2,915,576	December 1959
P 12	M. Strauss	US 2,905,739	September 1959
P 13	W. Krebs	US 3,108,908	October 1963
P 14	M. Mandel	US 2,941,022	June 1960
P 15	P. Ruetschi	US 2,951,106	August 1960
P 16	E. Voss	US 3,096,215	July 1963
P 17	F. Backmann	US 3,117,033	January 1964
P 18	P. Ruetschi	US 3,080,440	March 1963
P 19	P. Grieger	US 3,022,363	February 1962
P 20	J. Daley	US 2,980,747	April 1961
P 21	W. Garten	US 3,089,913	May 1963
P 22	A. Dassler	US 2,934,581	April 1960
P 23	W. Smith	US 3,057,942	October 1962
P 24	P. Grieger	US 3,037,066	May 1962
P 25	G. Neuman	US 2,934,580	April 1960
P 26	P. Denes	US 2,942,050	June 1960

P 27	R. Bloch	US 2,566,114	August 1951
P 28	W. Smith	US 2,588,170	March 1952
P 29	J. Moulton	US 2,634,303	April 1953
P 30	A. Warner	US 2,663,744	December 1953
P 31	L. Minnick *et al.*	US 2,863,933	December 1958
P 32	L. Minnick *et al.*	US 2,937,219	May 1960
P 33	A. Bopp	US 2,988,583	June 1961
P 34	J. Story	US 3,006,980	October 1961
P 35	J. Story	US 3,121,028	February 1964
P 36	M. Coler *et al.*	US 3,023,259	February 1962
P 37	D. Herbert *et al.*	US 3,043,896	July 1962
P 38	M. Lasser *et al.*	US 3,113,047	December 1963
P 39	S. Zaromb	US 3,114,658	December 1963
P 40	J. Duddy.	US 3,003,012	October 1961
P 41	M. Vogt	US 2,991,324	July 1961
P 42	K. Lehoves *et al.*	US 2,954,417	September 1960
P 43	F. Philipp *et al.*	US 3,003,014	October 1961
P 44	G. Drengler *et al.*	US 3,087,003	April 1963
P 45	M. Coler *et al.*	US 3,023,260	February 1952
P 46	L. Belone	US 3,083,249	March 1963
P 47	J. Duddy	US 3,121,029	February 1964
P 48	R. Jeannin	US 2,646,455	July 1953
P 49	F. Bonner	US 2,928,889	March 1960
P 50	E. Kujas	US 3,051,768	August 1962
P 51	M. Fukida	US 3,041,338	June 1962
P 52	E. Casey *et al.*	US 3,068,310	December 1962
P 53	G. Hartman *et al.*	US 2,865,973	December 1958
P 54	H. Vogt	US 2,627,531	February 1953
P 55	H. Vogt	US 2,681,375	June 1954
P 56	H. Vogt	US 2,836,641	May 1958
P 57	F. Peters	US 2,830,108	April 1958
P 58	J. Moulton	US 2,615,930	October 1952
P 59	L. Pucher *et al.*	US 2,850,555	September 1958
P 60	C. Freas	US 2,849,519	August 1958
P 61	F. Solomon	US 2,818,462	December 1957
P 62	R. Scheuerle *et al.*	US 2,862,985	December 1958
P 63	J. Duddy	US 2,881,238	April 1959
P 64	J. Duddy	US 3,007,991	November 1961
P 65	S. Corren	US 3,009,978	November 1961
P 66	E. Adler	US 3,024,296	March 1962
P 67	J. Eisen	US 2,708,683	May 1955
P 68	J. Eisen	US 2,902,530	September 1959
P 69	J. Duddy	US 3,121,029	February 1964
P 70	L. Urry	US 3,060,254	October 1962
P 71	R. Roberts	US 2,448,052	August 1948
P 72	R. Roberts	US 2,519,399	August 1950
P 73	F. Solomon	US 2,994,729	August 1961
P 74	A. Louis *et al.*	US 3,023,261	February 1962

P 75	J. Salauze	US 2,683,181	July 1954
P 76	J. Salauze	US 2,683,182	July 1954
P 77	J. Salauze	US 2,554,125	May 1951
P 78	J. Salauze	US 2,820,077	January 1958
P 79	J. Salauze	US 2,820,078	January 1958
P 80	J. Salauze	US 2,643,276	June 1953
P 81	J. Moulton	US 2,727,080	December 1955
P 82	J. Moulton	US 2,870,234	January 1959
P 83	J. Salkind	US 3,062,908	November 1962
P 84	L. Pucher *et al.*	US 2,833,845	May 1958
P 85	S. Ruben	US 2,554,504	May 1951
P 86	S. Dickfeldt *et al.*	US 3,076,860	February 1963
P 87	P. Haayman *et al.*	US 2,540,446	February 1951
P 88	J. Baldwin	US 2,653,179	September 1953
P 89	M. Yardney	US 2,983,777	May 1961
P 90	P. Ruetschi	US 3,057,444	October 1962
P 91	C. Chapman *et al.*	US 2,945,078	July 1960
P 92	V. Weil *et al.*	US 2,677,713	May 1954
P 93	J. Koerner *et al.*	US 2,728,808	December 1955
P 94	J. Dittman	US 2,866,840	December 1958
P 95	A. Sabatino *et al.*	US 3,100,162	August 1963
P 96	L. Scheichl	US 2,920,128	January 1960
P 97	C. Gritman *et al.*	US 2,715,151	August 1955
P 98	E. Sundberg *et al.*	US 2,853,536	September 1958
P 99	R. Greenberg	US 2,759,036	August 1956
P 100	R. Greenberg	US 2,759,037	August 1956
P 101	B. Cahan	US 3,017,448	June 1962
		Re 25,608	January 1964
P 102	J. Moulton	US 2,644,022	June 1953
P 103	J. Moulton	US 2,871,281	January 1959
P 104	H. Hignett	US 2,653,180	September 1953
P 105	L. Schlecht	US 2,700,062	January 1955
P 106	L. Schlecht *et al.*	US 2,699,458	January 1955
P 107	P. Jacquier	US 2,646,456	July 1953
P 108	P. Bourgaul *et al.*	US 2,831,044	April 1958
P 109	J. Salauze	US 2,833,847	May 1958
P 110	S. Corren *et al.*	US 3,009,979	November 1961
P 111	L. Goldenberg *et al.*	US 3,036,141	May 1962
P 112	L. Goldenberg *et al.*	US 3,036,142	May 1962
P 113	I. Blake *et al.*	US 2,534,403	December 1950
P 114	D. Louzos	US 2,535,742	December 1950
P 115	R. Kirk *et al.*	US 2,621,220	December 1952
P 116	J. Rhyne	US 2,828,350	March 1958
P 117	C. Morehouse *et al.*	US 2,809,225	October 1957
P 118	O. Lucas	US 2,706,213	April 1955
P 119	C. Morehouse *et al.*	US 2,759,986	August 1956
P 120	J. Cohn *et al.*	US 3,080,444	March 1963
P 121	M. Neipert *et al.*	US 3,134,698	May 1964

P 122	D. Sharpe	US 3,005,864	October 1961
P 123	H. Haring	US 3,007,993	November 1961
P 124	A. Warner *et al.*	US 2,661,388	December 1953
P 125	L. Pucher *et al.*	US 2,640,090	May 1953
P 126	A. Fischbach *et al.*	US 2,636,060	April 1953
P 127	M. Chubb *et al.*	US 2,715,652	August 1955
P 128	G. Ellis	US 2,655,551	October 1953
P 129	M. Chubb	US 2,658,935	November 1953
P 130	J. Salanze	US 2,744,948	May 1956
P 131	J. White *et al.*	US 2,492,206	December 1949
P 132	R. Reid	US 2,616,940	November 1952
P 133	K. LeLovec	US 2,689,876	September 1954
P 134	K. Barnard	US 2,732,417	January 1956
P 135	C. Morehouse	US 2,874,204	February 1959
P 136	M. Schwarz *et al.*	US 2,895,000	July 1959
P 137	S. Ruben	US 2,954,418	September 1960
P 138	J. Weininger	US 3,003,017	October 1961
P 139	F. Portail	US 2,840,626	June 1958
P 140	E. Schumacher *et al.*	US 2,848,525	August 1958
P 141	R. Glicksman *et al.*	US 2,897,249	July 1959
P 142	R. Blue *et al.*	US 3,019,279	January 1962
P 143	J. Booe	US 2,538,078	January 1961
P 144	J. Robinson	US 3,038,019	June 1962
P 145	M. Chubb *et al.*	US 2,715,652	August 1955
P 146	A. Fry *et al.*	US 2,547,907	April 1951
P 147	H. DeLong *et al.*	US 2,481,204	September 1944
P 148	I. Blake	US 2,612,534	September 1952
P 149	S. Ruben	US 2,948,769	August 1960
P 150	A. Fry *et al.*	US 2,547,908	April 1951
P 151	S. Bartosh *et al.*	US 3,025,336	March 1962
P 152	S. Ruben	US 2,948,768	August 1960
P 153	A. Fry *et al.*	US 2,712,564	July 1955
P 154	J. Stevens	US 2,934,583	April 1960
P 155	S. Ruben	US 3,048,645	August 1962
P 156	E. Anderson *et al.*	US 2,530,757	November 1950
P 157	P. George *et al.*	US 2,547,909	April 1951
P 158	D. Sargent	US 2,554,447	May 1951
P 159	J. Stokes	US 2,838,591	June 1958
P 160	J. Stokes	US 2,796,456	June 1957
P 161	S. Ruben	US 2,783,292	February 1957
P 162	J. McCallum *et al.*	US 2,979,553	April 1961
P 163	J. McCallum *et al.*	US 3,093,512	June 1963
P 164	J. McCallum	US 3,093,513	June 1963
P 165	J. McCallum	US 3,093,514	June 1963
P 166	J. Rhyne	US 2,837,590	June 1958
P 167	J. McCallum *et al.*	US 2,986,592	May 1961
P 168	S. Ruben	US 2,669,597	February 1954
P 169	J. Rhyne	US 2,837,591	June 1958

P 170	H. Hughes et al.	US 3,026,365	March 1962
P 171	T. Boswell	US 2,959,631	November 1960
P 172	R. Dean	US 2,605,297	July 1952
P 173	J. Moulton	US 2,964,582	December 1960
P 174	W. Krebs	US 3,055,963	September 1962
P 175	L. Urry	US 3,033,909	May 1962
P 176	J. Brennan	US 2,610,220	September 1952
P 177	M. Powers	US 2,428,470	October 1947
P 178	H. Zohn	US 2,694,099	November 1954
P 179	H. Zohn	US 2,694,100	November 1954
P 180	A. Brobe	US 2,794,845	June 1957
P 181	J. Doyen	US 2,867,678	January 1959
P 182	K. Brown	US 2,906,803	September 1959
P 183	F. Solomon et al.	US 3,055,964	September 1962
P 184	S. Ruben	US 3,066,179	November 1962
P 185	H. Zohn	US 2,953,619	September 1960
P 186	J. Lander et al.	US 2,821,565	January 1950
P 187	H. Stoertz	US 2,678,340	May 1954
P 188	C. Chapman	US 2,776,331	January 1957
P 189	J. Coleman et al.	US 3,081,369	March 1963
P 190	G. Heise et al.	US 2,612,532	September 1952
P 191	F. Schoeppe et al.	US 3,138,489	June 1964
P 192	H. Zimmerman et al.	US 2,572,296	October 1951
P 193	E. Schumacher	US 2,610,985	September 1952
P 194	M. Hatfield	US 2,615,931	October 1952
P 195	W. Darland et al.	US 2,914,595	November 1959
P 196	F. Schoeppe et al.	US 3,138,489	June 1964
P 197	J. Miller et al.	US 3,043,898	July 1962
P 198	F. Kordesch	US 2,935,547	May 1960
P 199	F. Everett	US 2,656,401	October 1953
P 200	F. Everett	US 2,901,525	April 1959
P 201	F. Everett	US 2,918,514	December 1959
P 202	H. Lawson	US 2,918,515	December 1959
P 203	A. Pitt	US 2,534,056	December 1950
P 204	J. Freund	US 2,981,778	April 1961
P 205	J. Freund	US 2,981,779	April 1961
P 206	T. Burnette	US 2,981,780	April 1961
P 207	H. Morton	US 2,996,564	August 1961
P 208	P. Williams	US 2,898,394	August 1959
P 209	E. Burrell	US 2,927,145	March 1960
P 210	J. Kuck	US 2,979,552	April 1961
P 211	H. Porter	US 2,682,567	June 1954
P 212	E. Burrell	US 2,931,848	April 1960
P 213	J. Jacobs	US 2,989,576	June 1961
P 214	H. Pattin	US 3,150,009	September 1964
P 215	C. Cleveland	US 3,150,010	September 1964
P 216	R. Barnett	US 2,990,442	June 1961
P 217	C. Cleveland	US 3,131,094	April 1964

P 218	L. Mott-Smith	US 2,921,974	January 1960
P 219	W. Darland *et al.*	US 2,985,702	May 1961
P 220	A. Saurwein	US 3,022,365	February 1962
P 221	N. Wales	US 2,403,567	July 1946
P 222	J. Henry	US 3,052,744	September 1962
P 223	R. Johnson *et al.*	US 2,945,079	July 1960
P 224	R. Johnson *et al.*	US 2,948,767	August 1960
P 225	W. Meyers	US 2,992,289	July 1961
P 226	V. Bryant	US 3,009,007	November 1961
P 227	W. Meyers	US 3,083,252	March 1963
P 228	M. Feinleib *et al.*	US 2,902,531	September 1959
P 229	W. Kearsley *et al.*	US 2,472,376	June 1949
P 230	L. Harriss	US 2,543,106	February 1951
P 231	J. Mullen	US 2,564,495	August 1951
P 232	J. Davis	US 2,594,879	April 1952
P 233	M. Comanor *et al.*	US 3,103,452	September 1963
P 234	J. Davis	US 2,634,305	April 1953
P 235	L. Pucher	US 2,667,527	January 1954
P 236	M. Chubb	US 2,684,395	July 1954
P 237	M. Chubb *et al.*	US 2,745,892	May 1956
P 238	M. Wilke	US 2,711,437	June 1955
P 239	W. Grupe *et al.*	US 2,806,077	September 1957
P 240	J. Dines	US 2,806,895	September 1957
P 241	H. McDonald	US 2,829,187	April 1958
P 242	J. Coleman *et al.*	US 2,829,188	April 1958
P 243	R. Jeannin	US 2,847,494	August 1958
P 244	E. Broglio	US 2,900,432	August 1959
P 245	H. Haring	US 2,988,587	June 1961
P 246	M. Wilke *et al.*	US 3,061,659	October 1962
P 247	F. Solomon *et al.*	US 3,100,164	August 1963
P 248	J. Pawlak	US 3,079,456	February 1963
P 249	H. Lawson	US 2,428,850	October 1947
P 250	H. Riggs *et al.*	US 2,404,144	July 1946
P 251	E. Barrett	US 3,017,449	January 1962
P 252	H. Heinsohn *et al.*	US 3,067,274	December 1962
P 253	H. Bauman	US 2,937,220	May 1960
P 254	J. Doyan	US 3,075,034	January 1963
P 255	L. Kardoff *et al.*	US 3,075,035	January 1963
P 256	M. Murphy	US 2,529,511	November 1950
P 257	E. Barrett	US 2,763,706	September 1956
P 258	C. Gold	US 2,783,291	February 1957
P 259	H. Recopé deTilly Blaru	US 2,787,650	April 1957
P 260	A. Renke	US 2,798,111	July 1957
P 261	J. Salange	US 2,852,592	September 1958
P 262	H. Recopé deTilly Blaru	US 2,862,038	November 1958
P 263	E. Smith *et al.*	US 2,905,741	September 1959
P 264	E. Cooper *et al.*	US 3,018,314	January 1962
P 265	J. Hopkins	US 3,053,928	September 1962

P 266	A. Fischbach *et al.*	US 2,679,457	May 1954
P 267	L. Urry	US 2,970,180	January 1961
P 268	J. Schrodt *et al.*	US 2,936,327	May 1960
P 269	H. Lawson	US 2,472,379	June 1949
P 270	S. Eidensohn	US 2,855,453	October 1958
P 271	A. Mansoff	US 2,626,971	January 1953
P 272	N. Kramer	US 2,761,006	August 1956
P 273	H. Riggs	US 2,418,790	April 1947
P 274	C. Endress	US 2,516,048	July 1950
P 275	J. Schrodt *et al.*	US 2,442,380	June 1948
P 276	G. Rezek *et al.*	US 2,679,549	May 1954
P 277	H. Weigand	US 2,938,066	May 1960
P 278	J. Akerman	US 2,700,064	January 1955
P 279	L. Lighton	US 2,410,952	November 1946
P 280	E. Martin *et al.*	US 2,666,091	January 1954
P 281	D. Carlson *et al.*	US 2,615,933	October 1952
P 282	H. Bauman	US 2,824,164	February 1958
P 283	M. Chubb *et al.*	US 3,022,364	February 1962
P 284	M. Comanor *et al.*	US 3,113,891	December 1963
P 285	G. DeMaio *et al.*	US 2,925,358	February 1960
P 286	R. Coolidge	US 2,683,102	July 1954
P 287	E. Otto	US 2,569,491	October 1951
P 288	M. Wilke	US 2,403,571	July 1946
P 289	G. Jobe	US 2,773,786	December 1956
P 290	H. Lawson	US 2,445,306	July 1948
P 291	D. Craig *et al.*	US 2,950,999	August 1960
P 292	H. Sperber	US 2,946,707	July 1960
P 293	H. Brown	US 2,857,295	October 1958
P 294	T. Johnson *et al.*	US 2,941,909	June 1960
P 295	M. Mendelsohn *et al.*	US 2,994,625	August 1961
P 296	M. Mendelsohn *et al.*	US 2,872,362	February 1959
P 297	S. Ruben	US 2,576,266	November 1951
P 298	H. Zimmerman *et al.*	US 2,900,434	August 1959
P 299	H. Zimmerman *et al.*	US 2,971,044	February 1961
P 300	K. Kordesch	Serial No. 689,083 filed	October 1957
P 301	K. Kordesch	US 2,991,325	July 1961
P 302	J. Fafa *et al.*	US 2,964,583	December 1960
P 303	J. Bikerman	US 3,137,594	June 1964
P 304	M. Friedman	US 2,600,526	June 1952
P 305	E. Voss	US 2,982,806	May 1961
P 306	H. Strauss	US 2,692,904	October 1954
P 307	H. Strauss	US 3,053,924	September 1962
P 308	H. André	US 2,594,709	April 1952
P 309	H. André	US 2,932,680	April 1960
P 310	P. Howard	US 2,724,734	November 1955
P 311	L. Ducher *et al.*	US 2,880,258	March 1959
P 312	R. Scheuerle *et al.*	US 2,865,974	December 1958
P 313	P. Garine	US 2,838,590	June 1958

P 314	W. Cunningham *et al.*	US 2,931,846	April 1960
P 315	H. André	US 2,988,586	June 1961
P 316	J. Duddy	US 3,003,015	October 1961
P 317	A. Fischbach	US 2,811,572	October 1957
P 318	H. André	US 3,082,279	March 1963
P 319	H. Strauss *et al.*	US 2,771,500	November 1956
P 320	M. Chubb *et al.*	US 2,727,082	December 1955
P 321	E. Hollman *et al.*	US 2,727,083	December 1955
P 322	W. Gary	US 2,669,595	February 1954
P 323	W. Schilke	US 3,123,504	March 1964
P 324	S. Corren *et al.*	US 3,009,980	November 1961
P 325	J. Daley	US 2,976,341	March 1961
P 326	E. Barrett	US 2,886,620	May 1959
P 327	H. André	US 2,594,714	April 1952
P 328	O. Jansson	US 2,702,308	February 1955
P 329	M. Stumbock	US 2,602,826	July 1952
P 330	K. Lindstrom	US 2,857,447	October 1958
P 331	H. Strauss	US 2,862,986	December 1958
P 332	K. Kordesch	US 3,042,732	July 1962
P 333	E. Jönsson *et al.*	US 3,049,578	August 1962
P 334	M. Chubb	US 2,817,697	December 1957
P 335	I. Blake	US 2,612,533	September 1952
P 336	A. Daniel	US 2,480,839	September 1949
P 337	R. Lewis *et al.*	US 2,571,732	October 1951
P 338	B. King	US 2,593,893	April 1952
P 339	A. Grusell *et al.*	US 2,586,426	February 1952
P 340	S. Ruben	US 2,620,368	December 1952
P 341	J. West *et al.*	US 2,723,301	November 1955
P 342	E. Leger	US 2,993,947	July 1961
P 343	M. Chubb *et al.*	US 2,852,593	September 1958
P 344	K. Jones *et al.*	US 2,939,900	June 1960
P 345	I. Denison *et al.*	US 2,513,292	July 1950
P 346	J. Moulton *et al.*	US 2,561,943	July 1951
P 347	I. Blake	US 2,612,536	September 1952
P 348	I. Blake	US 2,612,535	September 1952
P 349	I. Blake	US 2,612,537	September 1952
P 350	E. Klopp	US 2,897,250	July 1959
P 351	J. Davis	US 3,095,331	June 1963
P 352	S. Ruben	US 2,542,574	February 1951
P 353	S. Ruben	US 2,629,758	February 1953
P 354	W. Taylor	US 2,647,938	August 1953
P 355	S. Ruben	US 2,536,699	January 1951
P 356	S. Ruben	US 2,542,710	February 1951
P 357	S. Ruben	US 2,462,998 Re 23427	March 1949
P 358	S. Ruben	US 2,463,316	March 1949
P 359	S. Ruben	US 2,499,419	March 1950
P 360	S. Ruben	US 2,536,696	January 1951

P 361	S. Ruben	US 2,481,539	September 1949
P 362	J. Boswell	US 2,683,184	July 1954
P 363	J. Boswell	US 2,959,631	November 1960
P 364	H. Hughes *et al.*	US 3,026,365	March 1960
P 365	T. Moir	US 2,441,896	May 1948
P 366	A. Bapp	US 2,988,583	June 1961
P 367	O. Reinhardt	US 2,519,053	August 1950
P 368	W. Woodring	US 2,519,054	August 1950
P 369	J. Marinez *et al.*	US 2,521,800	September 1950
P 370	S. Ruben	US 2,487,985	November 1949
P 371	W. Woodring	US 2,526,789	October 1950
P 372	L. Quinnell	US 2,525,270	October 1950
P 373	S. Ruben	US 2,526,692	October 1950
P 374	E. Klopp	US 2,923,757	February 1960
P 375	H. André	US 3,075,032	January 1963
P 376	H. André	US 2,669,594	February 1954
P 377	M. Yardney *et al.*	US 2,635,127	April 1953
P 378	H. André	US 2,594,710	April 1952
P 379	H. André	US 2,594,711	April 1952
P 380	H. André	US 2,594,712	April 1952
P 381	M. Yardney	US 2,610,219	September 1952
P 382	C. Horowitz *et al.*	US 3,098,770	July 1963
P 383	J. Coleman *et al.*	US 2,597,451	May 1952
P 384	J. Coleman *et al.*	US 2,597,452	May 1952
P 385	J. Coleman *et al.*	US 2,597,453	May 1952
P 386	J. Coleman *et al.*	US 2,597,454	May 1952
P 387	J. Coleman *et al.*	US 2,597,455	May 1952
P 388	J. Coleman *et al.*	US 2,597,456	May 1952
P 389	D. Herbert *et al.*	US 3,043,896	July 1962
P 390	L. Minnick *et al.*	US 2,863,933	December 1958
P 391	E. Beer	US 2,752,550	June 1956
P 392	C. Ferguson *et al.*	US 2,935,675	May 1960
P 393	V. Romanov	USSR 140,835 (CA **56**: 9882f)	September 1961
P 394	S. Mayer *et al.*	US 3,185,590	May 1965
P 395	G. Mueller	US 3,184,338	May 1965
P 396	B. Bartfai *et al.*	US 3,185,591	May 1965
P 397	L. Horn *et al.*	US 3,180,761	April 1965
P 398	E. Voss	US 2,982,806	May 1961
P 399	L. Borbachevskaya *et al.*	USSR 160,207	January 1964
P 400	L. Borbachevskaya *et al.*	USSR 160,209	January 1964
P 401	H. André	US 3,184,340	May 1965
P 402	T. Buitkus	US 3,174,880	March 1965
P 403	C. Wolfenbüttel	Ger. 1,170,659	May 1964
P 404	M. Pryor *et al.*	US 3,189,486	June 1965
P 405	A. Schmid	Belg. 448,479 (CA **41**: 6166b)	January 1943
P 406	P. Dénes	Austrian 196,464	March 1958

P 407	P. Dénes	Israel 9023	March 1957
P 408	F. Bachman	Ger. 1,015,874	September 1957
P 409	T. Edison	US 1,016,874	February 1912
P 410	A. Dassler	US 2,104,973	January 1938
P 411	J. Watanabe et al.	Japan 1580	1957
P 412	Z. Sharunova et al.	USSR 149,479	August 1962
P 413	W. Vielstich	Ger. 1,118,843	December 1961
P 414		Ger. 1,141,689	December 1962
P 415	E. Beer	Dutch 73,483	October 1953
P 416	W. Bauer	US 1,134,093	April 1915
P 417	W. Arsen	US 2,306,927	December 1942
P 418	D. Sargent	US 2,554,447	May 1951
P 419	M. Mendelsohn	US 2,915,579	December 1959
P 420	R. Panzer	US 3,117,032	January 1964
P 421	J. Bone	US 2,740,821	April 1956
P 422	J. McCallum et al.	US 3,033,910	May 1962
P 423	P. Marsal et al.	US 2,977,401	March 1961
P 424	O. Brill et al.	US 2,654,795	October 1953
P 425	P. Ruetschi	US 3,160,526	December 1964
P 426	D. Stanimirovitch	US 2,740,824	April 1956
P 427	H. Honer	US 3,205,096	September 1965

Index

301